Sacred Landscapes of Imperial China

Giulio Magli

Sacred Landscapes
of Imperial China

Astronomy, Feng Shui, and the Mandate
of Heaven

 Springer

Giulio Magli
Department of Mathematics
Politecnico di Milano
Milan, Italy

ISBN 978-3-030-49326-4   ISBN 978-3-030-49324-0   (eBook)
https://doi.org/10.1007/978-3-030-49324-0

This Springer imprint is published by the registered company Springer Nature Switzerland AG
The registered company address is: Gewerbestrasse 11, 6330 Cham, Switzerland

The original version of this book was revised. Credits for images have been included in the Frontmatter.

Introduction

You look and you marvel.

A couple of life-sized, recumbent stone elephants greet you along the way, and as you look back and forth, dozens of other statues do the same. On the far horizon behind you, a prominent mountain occupies the scene in alignment with a huge, five-arched doorway which, however, has no walls at its sides. When you look forward, the path bends for no apparent reason, but you do know that it is because malevolent spirits only go straight, and that the magnificent tomb of the emperor will be at the end of the path, to the north, still many kilometers away. Beautiful, dragon-shaped hills harmoniously surround the place to the east and west, and a river quietly flows to the south.

Some readers might have recognized the place I am speaking about: It is the UNESCO site of the Necropolis of the Chinese emperors of the Qing Dynasty traditionally called the Qing Eastern tombs. We shall be visiting and studying this place in great depth in this book: The Qing Eastern Necropolis is indeed just one example of the many astounding masterpieces—arising from a skillful combination of man-made and natural features—to which this book is devoted: the sacred landscapes of the emperors of ancient China. Places of power, beauty, peace, and eternity that the Sons of Heaven devised over the course of two millennia according to complex rules which governed the choice of the sites, the architecture of the buildings, and their mutually related placement in the landscape.

Such rules were codified into doctrines: the *Zhao Mu* first and the *Feng Shui* later. Feng Shui, in particular, is a form of "geomancy": According to it, there exists a "vital energy"—the *Qi*—flowing on the Earth. Establishing auspicious places can be done by individuating the flow of the Qi on the basis of the morphology of the land and/or the direction of a magnetic needle. Of course, Physics tells us that there are no such things as "earth energy flows," and therefore, Feng Shui is a "pseudo-science," as Joseph Needham called it. I do not like this terminology however, because Feng Shui started to exist millennia before the thing it purportedly apes. My personal background is in Astrophysics and the viewpoint of this book is wholly scientific; however, I have a profound respect for the ideas and the minds of the ancient builders who conceived the architectural wonders we shall visit in this book, as well as for the sense of beauty associated with Feng Shui, which had an enormous influence on Chinese architecture. Our approach here is therefore similar to the one that must be used with Astrology when studying Archeoastronomy: Astrology is a superstition which has nothing to do with science, but taking into account the role of Astrology is in many cases fundamental to understand the connections of ancient buildings with the sky (for instance, the sign Capricorn played a crucial role in Augustus' iconography and building program). This analogy actually goes far beyond a theoretical similarity. Interestingly, we shall see that Feng Shui "alignments" can be studied using the same methods that modern Archeoastronomy employs for astronomical alignments. Furthermore, other important architectural canons in China came about as a result of the celestial connection of the Emperors with the Northern sky and were therefore governed by astronomy. Thus, Archeoastronomy will also come into play in this book in its proper sense. Finally, as far as alignments possibly governed by magnetic compass are concerned, to study them we shall make use of a few notions of Paleomagnetism (as far as the background required is concerned, the book is self-contained: the technical concepts needed for an in-depth reading are provided in the Appendix).

All in all, at least to the best of the author's knowledge, this book is the first existing attempt at a comprehensive study of the role of cognitive aspects—in particular, Astronomy and Feng Shui—in Chinese imperial architecture. However, this book is also a "cognitive voyage" of the discovery of spectacular monuments—including some universally known UNESCO sites, but also places virtually unknown to the general public—in search of the way of thinking and of the technical skills of their builders.

Spanning a time arc of more than 2100 years, we will thus follow the impressive cultural continuity of Imperial China in a way which—I hope—will help us to increase our understanding of this magnificent civilization.

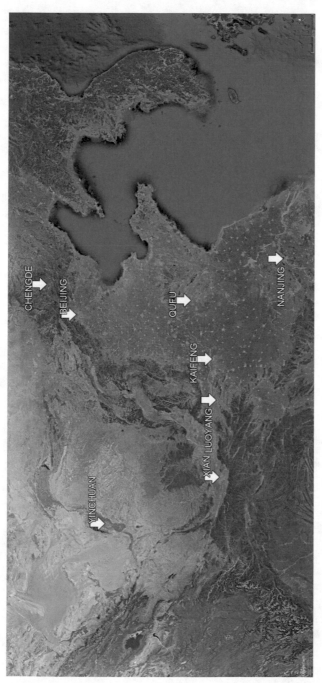

A general map of China, with the main cities mentioned in the book

Contents

List of Figure Credits

Chapter 1
Heavens and Earth in Ancient China

1.1 Religion and Natural Philosophy

The exact day is unknown. Yet, it is certain that 221 BC marks a fundamental break-through in China history. For it was in that year that the army of King Zheng of the state of Qin completed his conquest of the country, defeating the last of the inde-pendent reigns, the state of Qi (modern Shandong). Immediately afterwards, Zheng proclaimed himself as *Shihuang*, the First Emperor. It will be with the first Emperor and from his astonishing terracotta army that our journey into the sacred landscapes of ancient China will start (Chap. 3). First, however, it is necessary to equip ourselves with a certain amount of notions about the Chinese worldview, which formed over the course of the millennia preceding the Qin unification and later crystallized in that peculiar and complex mixture of natural philosophy, knowledge and beliefs which goes under the name of Feng Shui.

The earliest evidence of agriculture and sedentarism in China can be traced back to 8000 BC in the area of the Yangtze River, while later phases such as the Yangshao culture (5000–3000 BC), and the Longshan culture (3000–2000 BC) were centered on the Yellow River. Traditional historiography places the first Dynasty of Chinese rulers, the Xia, in the period 2000–1600 BC circa. The Xia Dynasty has been consid-ered mythical by many historians, but an increasing amount of evidence (e.g. from the excavations at Erlitou) points to the existence of a centralized state in precisely that period. The Shang Dynasty (1600–1046 BC) and the following Zhou dynasty have instead been definitively proved archaeologically. The Zhou dynasty formally runs up to the third century BC. In the eighth century BC however, local rulers began to accumulate enough power to declare themselves kings, and China entered a long and increasingly chaotic era, divided by historians into the Spring and Autumn (770–475 BC) and the Warring States (475–221 BC) periods.

The religious beliefs of the Shang and the Zhou included celestial (star) deities, terrestrial deities (like gods of the land, the mountains, and the rivers) and ancestor worship. These three categories of divine beings were actually to accompany the whole history of religion in China. During the late Zhou and Warring States periods, though, also 'subterranean', chtonic deities and functionaries of the Netherworld would appear, as well as the belief in dangerous ghosts resulting from unnatural death. Besides the persistence of this system of beliefs, at the end of the Spring and Autumn period and during the Warring States era, a series of ideas and doctrines were developed and amalgamated into a sophisticated philosophy of the cosmos, which was to profoundly influence the worldview of Imperial China for two millennia (Thote 2009; De Groot 1902).

The first of such doctrines, directly based on natural philosophy, was Taoism. Tao is the (divinized) harmony of the cyclical behavior of the universe, and a Taoist should seek to uphold and respect this, since human interactions with each other and with nature can cause perturbations in the proper flow of the universal order. Ethical values which should inform this road to perfection include compassion, frugality and humility, as taught in the most important text of early Taoism, the *Tao Te Ching*. This text is traditionally attributed to an author called Laozi, whose identity is uncertain, but who was probably active at the end of the fourth century BC. Since then, Taoism has been profoundly embedded in Chinese history and tradition. The Pantheon of Taoism is dominated by the abstract entity of the Tao but is open to a hierarchy of different levels of deities, including already ancient Gods such as the "Queen Mother of the West", an archaic Goddess associated with the western mountains, which Taoists will also associate with their view of immortality. Taoists indeed follow rituals and exercises, and take substances, aimed at aligning themselves spiritually with the cosmic force. This alignment should improve health and extend life and longevity, indeed to the point of—in principle——attaining immortality. This led the Taoists to put substances to trial and to develop alchemy and elixirs.

Almost contemporary and intertwined with Taoism, a parallel philosophical school developed based on a dualist conception of nature. Founded by Zou Yan (305–240 BC), the dualism of Yin and Yang opposites (where in general terms Yang is related to masculinity, light, roundness, heaven and Yin to femininity, dark, square-ness, earth, among others), leaving aside the trivial interpretations which persist today, was a philosophical attempt at a cosmic explanation of nature in terms of basic principles not much dissimilar to the contemporary attempts of the Greek philosophers. According to this school, the balance of the two opposites assured the order and the cyclical alternation of nature, itself made up of different phases of five elements: water, fire, wood, metal, and earth. The union of Yin and Yang forces was conceived of as the source of life, and this symbolic polarity was reflected in the Heavens: by the alternation of the sun and the moon, and by two legendary deities connected with them, Nuwa and Fuxi. In Han burial murals, for instance, the sun is generally represented by a black bird and the moon with a green toad, and the two supernatural creatures Nuwa and Fuxi (with a human face and a snake-dragon's body) are depicted embracing the sun and the moon and holding two architects' instruments: a square and a compass. These tools allude to the shaping of the universe and to the cosmic order.

The life of the most famous of the Chinese philosophers, Kong Fuzi (551–479 BC), universally known by the Latinized form of his name, Confucius, also belongs to the end of the Spring and Autumn period. Confucius' primary concern was the respect of the ethical values of human society: morality, propriety, sincerity. These tenets were set out in the *Classics*, a series of texts credited directly to Confucius (although modern critics are cautious) and in the famous *Analects*, compiled after his death. Confucius' ideas were further developed and spread by his disciples, especially Mencius and Xun Zi. The first Confucian loyalty was to the family: respect of elders and parents, veneration of ancestors. Respect is also due to the past in general, and this led Confucians to place special importance on the study and emulation of past examples. Politically, Confucius—living in an endless period of wars between competing states—invoked unification and a centralized government. In spite of this, Confucianism was too mild for the "legalistic" (absolutist) approach of Shihuang, but emerged as a sort of imperial state philosophy slightly later, with the Han dynasty, and was consolidated over the centuries as the cornerstone of Chinese society (Fields 1989). A final boost to the adoption of Confucianism as a state doctrine occurred, as we shall see later, during the Song dynasty when it merged with ideas of Taoism and Buddhism, generating so-called Neo-Confucianism. In particular, the famous tradition of Chinese imperial examinations was closely linked with the "official" aspect of Confucianism. The system, aimed at selecting candidates for the state administration, started with the Han and became progressively more important under the Tang and the Song dynasties. Successful candidates had to exhibit a perfect knowledge of calligraphy and of the classics; passing the exams could change one's life forever, but the examination was an horrific experience, as anyone who has visited the (still existing) examination cells of the Jiangnan Gongyuan, the Imperial Examination Museum of China in Nanjing, can attest. Finally, the question as to whether Confucianism is a true religion or not is delicate and will not be touched upon here; what is certain is that it is followed in a religious manner by many, and that ceremonies associated with sacrifice to ancestors and deities of various types are an important part of it. Respect for nature and for the natural cycles, and ideas on the afterlife, also arise at many points in Confucius' and Mencius' writings (Eno 1990).

A further, major influence on the Chinese conception of nature and environment is that of Buddhism, which came to China from India in the first century AD. The founding values of Buddhism, and in particular the respect for life in all its manifestations, added to the Chinese worldview, enhancing the idea that everything in the world has a relationship with everything else, be it animate or inanimate, and contributes to the regular flow of the Cosmos. Many important Emperors throughout the history of China espoused Buddhism: to cite just one of the examples we shall encounter in this book, one might mention the Tang Emperor Gaozong and his wife Wu Zetian, to whom we owe two masterpieces of Buddhist architecture: the Wild Goose Pagoda in Xi'an and the Longmen Grottoes in Luoyang.

All things considered, one might venture to say that all these doctrines, although remaining independent and alternately followed by different rulers also within the course of the same dynasty, collectively contributed to the creation of a comprehensive worldview that placed the search for harmony and the respect for the regularity of the natural cycles at center stage. The existence of a unifying concept which inspired the relationship between the man-made features of the landscape and the natural environment over the course of many centuries of Chinese history should not, therefore, come as a surprise: the concept of *Qi*.

Variously translated as "vital energy", "energy flow", "life force" and the like, it has been mentioned in various forms since the time of Confucius and other early writers, who refer to Qi as an interior spirit of individuals. Zhuangzi, one of the founders of Taoism, was the first to write about Qi as something which is pervasively expressed in natural phenomena (the wind for example) and can be accumulated and dispersed. Other texts of the 4-3 century BC explicitly assign the presence of Qi to all living things, ranging from trees and vegetation. Finally, in a collection of philosophical and moral essays—called *Hainanzi*—which was certainly composed before 139 BC, it is explicitly stated that "The universe produces Qi" and that the formation and transformations of Qi produce the different Yin and Yang aspects of nature: for instance hot Qi/Yang produces fire, cold Qi/Yin produces water, and so on.

1.2 Divination, Astrology and Astronomy

The complex philosophical ad religious view of nature which, as we have seen, accumulated in China in the first millennium BC was merged in the following centuries with another—complex, albeit quite diffuse—doctrine: divination. Subsumed under the general term "divination", a variety of very different practices are usually meant. Common to such practices is the claim to be able to establish the meaning or the causes of events, or to foretell the future, and/or to reveal the will of the gods. Divination practices are also aimed at establishing lucky or unlucky days for doing things, or establishing auspicious or inauspicious places and landscapes.

Divination in the ancient world took very different forms and was practiced using very different means, a famous case being the "reading" of sheep's entrails by Etruscan and Roman diviners. During the Shang period in China, divination took the form of the reading of "oracle bones". These are cattle bones or turtle carapaces which were heated to produce cracks, to then be interpreted by the diviners. Both the questions posed and the answers obtained were written on the bone, so that the finding of thousands of such items, occurring in the 1950s, constituted a substantial source of information. In particular, we know that divination was practiced at the court level and thus rulers relied on it to interpret the will of Heaven as an aid to taking decisions. Chinese divination methods, however, changed over time, as reading of the bone cracks was gradually replaced by other methods, based on throwing sticks or drawing lots and interpreting the results. At least from the third century BC, diviners also started to make use of an instrument—the *shih* or "diviner's board"—which consisted of two superimposed plates of bronze or painted wood (Harper 1978). The upper plate was a disc, while the lower was square in shape. The two shapes—circular, and squared—correspond in the Chinese tradition to Heaven and Earth respectively, and indeed the center of the circular, heavens' plate contained a representation of the constellation of the Dipper (the Great Bear). The upper part revolved on a central pivot and had engraved upon it 24 compass-points. The ground plate was marked, all around its edge, with the names of the 28 *hsiu* (divisions of the sky, see below), and the 24 directions were repeated along its inner gradations. Also present on the board were the symbols of the *Bagua*, the traditional eight symbols used in Taoist cosmology. Also called trigrams, they are made up of three lines, each line either broken or unbroken, giving a total of eight different combinations (fragmentary examples of divination boards are documented archaeologically from Han-period tombs).

Contemporaneous with all other forms of divination, astrology also had a place in China's courts from early times, and was used to make predictions about affairs of State. As usual in ancient times, astrology and astronomy evolved side by side: on the oracle bones, for instance, we can find records of eclipses and other astronomical phenomena coupled with divination texts. We refer the reader to specialized books for details of the history of Astronomy in China (see, in particular, Pankenier 2013; Needham 1959); I shall discuss here only a few general facts that will be important in what follows.

Astronomers were, first of all, in charge of maintaining an accurate calendar, since the rulers (as we shall see better in the sequel) identified themselves as the keepers of the regularity of the celestial order. Chinese astronomical texts of which we are aware go as far back as the Warring States period, but two very important documents come from a slightly later time. These are the texts which have been discovered—together with other texts of scientific interest, like cartographic maps—in the so-called Mawangdui tombs in Changsha, Hunan (Buck 1975; Hsu 1978; Zhongshu 1992; Stephenson 1994). These tombs are of fundamental importance for understanding Chinese burial customs and beliefs, and we shall encounter them again later on. One of the texts is an astrology book recording positions of planets, the other a sort of illustrated guidebook to different types (tails) of comets. Besides the dating of this material, which must be earlier than the tombs, we are sure that accurate observation of the sky must have commenced well before. For example, material evidence of early astronomical observations comes from Taosi, Shanxi, an important settlement dated to 2300–1900 BC. Here a terrace of rammed Earth with three concentric semi-circular levels has been unearthed. On the rims of the innermost terraces are the foundations of thirteen pillars which, when viewed from an observing point close to the center of the altar, could be used for accurate observations of the rising positions of the sun over the course of the year (He 2018; Wu et al. 2009).

To chart the sky, Chinese astronomers introduced accurate mapping of the celestial regions, based on twenty-eight *hsiu* ("mansions") and three "enclosures" (concentric regions) centered on the North Celestial Pole. The three enclosures were called the Purple Forbidden enclosure, the Supreme Palace and the Heavenly Market. The first is also the most important as it occupies the circle closest to the celestial pole. It is therefore the center of the heavens, circled by all the other stars; in western terms it essentially includes the constellations of the Northern sky Ursa Minor, Draco, Camelopardalis, Bootes and Ursa Major (of course, the Chinese way of grouping the stars—and therefore the constellations—was generally quite different from ours). This region of the sky was of paramount importance also because the function of the pole as pivot of the sky was equated with the centrality of power on Earth, in accordance with a synergism between astronomy and power, which was in play in several civilizations around the world (see e.g. Magli 2013). In China, this connection has been documented since at least the Shang Dynasty (mid-second millennium BC) if not before (Pankenier 2015). The north-pole—*beiji* or "Northern Culmen"—became a "celestial archetype" of the semi-divine nature of the monarchs, and even Confucius—though with his typical secular approach—was to use the regularity of the rotation of the stars around a fixed center as a paragon of virtuous conduct in the government of a country. Interestingly, at the very same time, the Taoists would identify the fixed point in the sky with the ultimate achievement of the Tao balance and harmony, not only for the individual but also for good government. It should be recalled at this point that we are accustomed to identifying the north celestial pole quite easily: it is sufficient to search for Polaris, the Pole star. However, even today, the pole *does not* coincide with Polaris, and in ancient times—in particular, at the end of the first millennium BC—it was much farther. There is, in fact, the phenomenon called *precession*. Precession is the rotation of earth's axis around the perpendicular to the Ecliptic (the plane which contains the sun and the earth's orbit). The axis completes a revolution every 25,776 years. Since the sun is on the Ecliptic, precession does not affect the apparent motion of our star as seen from the Earth. Taking into account precession is, however, essential for reconstructing the skies as ancient cultures saw them, since it changes—slowly, but inexorably—the declination (the position on parallel circles drawn in the sky) of all the other stars over the centuries. In particular, the north celestial pole is seen to move along the path traced in the sky by the extension of the Earth's axis: a circle. In a certain epoch, its position on this path may or may not be near a brilliant star. What we call the Pole Star (Polaris) is actually the star to which the north pole has been nearest to from some centuries, but in ancient times it was not so. Around 2800 BC, the star Thuban of the constellation Draco was a good Pole Star, but later the north celestial pole ventured into a dark region. As a matter of fact, in the first millennium BC, the pole did go relatively close (but always at a few degrees of distance) to a bright star, Kochab of the Little Dipper, while at the end of the millennium the pole started to move slowly towards "our" pole star, Polaris. The accuracy of Chinese astronomers leads us to think that—although a formal explanation of the physical causes of precession was only to occur in Chinese texts in the third century AD—they had been empirically aware of the drift of the north celestial pole in relation to the stars, and may even have attempted to trace the path of the pole through barely visible stars such as those of the constellation we call Camelopardalis. It is difficult to identify the time at which this awareness came about, but it must have been at least from the first half of the first millennium BC. Their discovery was facilitated by the fact that they devised a method of measuring what they called "the four excursions"—upper and lower culmination, east and west elongation—of a circumpolar star using a sighting

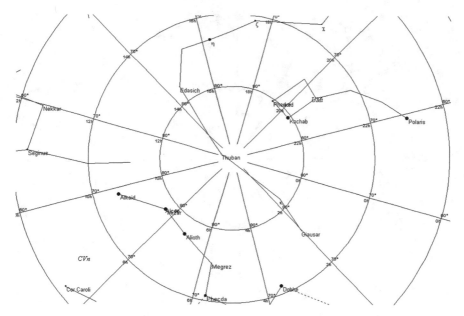

Fig. 1.1 The position of the north celestial pole relative to the stars around 2800 BC: There is a "pole star", Thuban

tube (of course with no lenses) with indentations, which must have been capable of discovering the precessional drift and estimating its behavior, at least in the near future (Needham 1959) (Figs. 1.1, 1.2, 1.3).

Among the stars of the northern region, the Big Dipper (the seven brilliant stars of the constellation we call Ursa Major) played a special role. Its rotation was used to mark the seasons of the year: indeed a characteristic of the Big Dipper and of all others circumpolar stars is that—contrary to the stars which rise and set, which have a period of conjunction with the Sun and therefore a period of invisibility—they are visible every night, but in different quarters of the sky with respect to the pole, depending on the season. With the Han dynasty, explicit representations of the connection of divine power with the northern sky and in particular with the Big Dipper occur. Of special relevance in this context are the reliefs of the funerary shrines built in honor of Wu Liang (died in I51 AD) and of two of his relatives in Shandong. The complex iconography of these reliefs gives us many information about the Han culture and society, and can be divided into three main themes: examples of correct behavior and filial piety (in accord with the main values of Confucianism), the afterworld and the journey of the soul, and finally the Heavens and the good omens as signs of Heaven's response to virtuous conduct. In this context the God of the Northern Sky *Di* is shown, dressed as a mythical emperor and driving a celestial carriage depicted as connecting the seven stars of the Dipper (Fairbank 1941; Drake 1943; James 1988; Powers 1991) (Fig. 1.4).

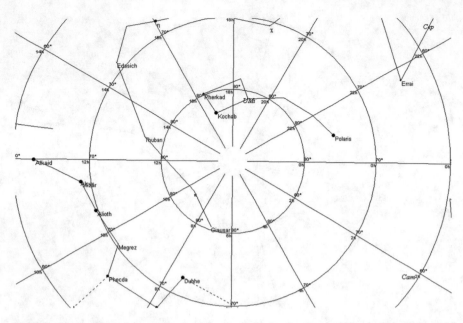

Fig. 1.2 The position of the north celestial pole relative to the stars in the early Chinese imperial times: There is no pole star; the pole has not been far from Kochab and is now moving towards Polaris

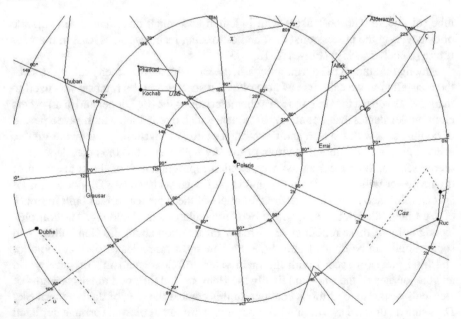

Fig. 1.3 The position of the north celestial pole relative to the stars today: Polaris is the pole star

Fig. 1.4 Rubbing of a stone relief, Han Dynasty. The Celestial Thearch is driving his heavenly chariot, the Big Dipper. *Credits* Images in the public domain

From the Han period we also have definitive texts confirming this celestial identification of the Emperor—who, not by chance, starts to be called Di as well. The most important of these texts is the one to be found in the astronomical-astrological part of the *Shiji*—the *Records of the Grand Historian*—the famous treatise written by Sima Qian (145–86 BC), the father of Chinese historiography, a personage we will also encounter later on. Sima had a tragic life: he inherited his father's position as court historian in the Han imperial court, but was persecuted for speaking in defense of a general blamed for an unsuccessful campaign. Having being given the choice of being executed or castrated, he chose the latter, having in this way the opportunity of finishing his work. His court position involved the mastery of the knowledge of Astronomy and Astrology, and before the tragic events, Sima Qian was involved in the creation (104 BC) of a reformed calendar, well-anchored with the solar year, allowing his Emperor Wu to change his era name as "supreme beginning" (see Sect. 4.3). According to Sima (Pankenier 2005):

> The Dipper is the Celestial Thearch Di's carriage. It revolves about the center, visiting and regulating each of the four regions. It divides yin from yang, establishes the four seasons, equalizes the Five Elemental Forces, deploys the seasonal junctures and angular measures, and determines the various periodicities: all these are tied to the Dipper.

The identification of the emperor with the north had important consequences for ceremonial orientation: the living emperor usually faced south in audiences and on court occasions, so that people prostrated themselves towards the north. Further, the dead emperor formally overlooked visitors to his tomb from the north, so that—generally speaking—burial mounds and tombs' access ways were oriented south-to-north (specific orientations and their symbolic reasons will be discussed at length in the sequel).

The second enclosure of the Chinese astronomers, the Supreme Palace, corresponded essentially to our constellations Virgo, Coma Berenices and Leo, while the third, the Heavenly Market, covered the constellations Serpens, Ophiuchus, Aquila and Corona Borealis. Besides these three enclosures, the remaining regions of the sky were mapped with the so-called Twenty-Eight Mansions. They were grouped into Four Symbols of seven mansions each, each symbol being associated with a compass direction. The symbols correspond to mythical animals; their names, and the main western constellations corresponding to each group, are: the Azure Dragon of the East (Libra and Scorpio) the Black Tortoise of the north (Capricorn, Aquarius, Pegasus), the White Tiger of the West (Aries, Taurus, Orion) the Vermilion Bird of the South (Cancer, Hydra). Intriguingly, it appears that the tradition of the "directional animals" is extremely old. Indeed, in a tomb located in Xishuipo, a Neolithic site in Puyang, Henan, associated with the Yangshao culture, a very curious discovery has been made. In the tomb, the body of a tall adult male was oriented south-north and flanked by two mosaic images made up of white clam shells; a third figure comprising a mosaic triangle and two human tibia was placed at his feet. The two images in the mosaics are easily recognizable as a tiger to the west and a dragon to the east of the body. The third image—in the north—might be a representation of the Big Dipper. Dating from the late 4th millennium BC, this grave certainly belongs to a chief, or perhaps a Shaman. It is difficult to establish to what extent it shows a "cosmographic" vision of the world, but the fact that the two animal beings—the Tiger and the Dragon—were associated with the cardinal directions west and east respectively so far back in antiquity is undeniable (Fig. 1.5).

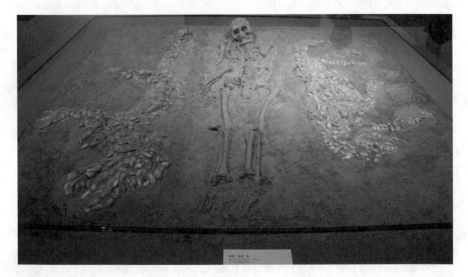

Fig. 1.5 Burial site from Xishuipo, Puyang, Henan. The body of a tall adult male (head originally towards the south) is flanked by two mosaics of white clam shells: A tiger to the right (west) and a dragon to the left (east)

1.3 From Natural to Sacred Landscapes

The Chinese philosophy of Nature—taken in the broadest sense and therefore including knowledge of the celestial and natural cycles as well as the religious interpretation of them—influenced the development of the Chinese landscapes through a series of cognitive aspects which can be better understood by making use of a concept—that of *sacred space* or *sacred landscape*—originating in seminal studies by the historian of religion, Mircea Eliade (1959, 1964, 1971). Eliade was the first to understand a series of mechanisms of the sacred which furnish a powerful key to cognitive aspects of religion. In particular, he wrote:

> The experience of sacred space makes possible the founding of the world: where the sacred manifests itself in space, the real unveils itself, the world comes into existence.

Thus a sacred space is a place "where the sacred manifests itself". Typically then, a space becomes sacred if people are able to recognize in it their religious worldview, or "cosmovision". But any vision of the Cosmos includes as a key ingredient the constant and regular renewal of the natural cycles, and this in turn brings into play the concept of order. Thus space becomes sacred space—a place where the worldview is manifest—when it is ordered through "cosmization". Cosmization was itself closely linked with the sky: indeed, time is ordered if framed in a calendrical scheme, and this derives from the observation of the celestial cycles; space is ordered if the cardinal directions are identified, and this is done by using the apparent motion of the stars around the celestial pole and/or the daily, apparent motion of the Sun.

This general description of sacred space can be applied to countless many cases, from the Necropolis of the IV Egyptian dynasty on the Giza Plateau to the place that Augustus chose to celebrate his new political order, the Campus Martius in Rome and its main monument, the Pantheon (Hannah and Magli 2011). Not surprisingly, in these examples and in many others, the afterlife and the celebration of the ancestors have a prominent place: the contact with the divine and the supernatural is indeed a structural component of any worldview, as it helps in establishing a cultural memory, in which the accepted cosmovision is accommodated and kept alive. The person in charge of communicating with the "outer world" is usually identified in anthropological terms as the *Shaman*. Shamanistic functions can be the duties of specialized figures, or can coincide with the divine powers of the rulers themselves, as in China, where the Emperors credited themselves with a mandate from Heaven, or in Egypt, where the Pharaohs were considered directly as living Gods. In any case, keeping alive the established view is fundamental for the stability of the power which is based on it. This is usually done by rituals of renewal, but *must be done in appropriate places*. This is the premise that led to monumental architecture, built in sacred spaces, created to act as explicit witness to the legitimacy of power. Monumental architecture thus stemmed from—and was a symbol of—power, and as such—again—it had to be explicitly connected with the cycles of the cosmos. The mechanism for achieving this was defined by Eliade as the *hierophany*. A hierophany (from a combination of Greek words, meaning "the sacred reveals itself") is an explicit manifestation of the sacred: of a god, of a message from a god, or even of a divine aspect of nature. Instances of hierophanies are just the various means of communication with divine power (for example, official festivals in which the power of the gods is symbolically renewed). However, hierophanies go far beyond this. As we have seen, space is validated and has to *deserve* to be lived in, if and only if it conforms to patterns and rules determined by the sacred. In a sense, then, the only true space is the sacred space. But if the space is sacred, then it is also a place where the divine can—or even must—manifest itself. So, a hierophany can be a truly physical, tangible manifestation of the sacred in a place that is worthy of allowing such a manifestation to occur. Hierophanies can appear as abrupt flashes: one key example is the working of astronomical alignments, in which—for instance—the sun rises/sets at the prescribed point on the day it is designed to do so. But a hierophany can also be *perennial*: a image, or a set of images, which is visible every day and conveys a message of contact with the sacred. For the present book, perennial hierophanies are the most important. Let us, therefore, consider a few examples.

A first example comes from the Mayas, the Mesoamerican civilization which developed in present-day southeastern Mexico, Guatemala, Belize, and Honduras. During the so-called Classic period (250–850 AD circa) in the Yucatan (not far from the Mayan city of Chichen Itza') the Mayan priests discovered a cave, called Balancanché, which became a center of cult of the god of rain. The cave was walled up by the Mayas themselves at the end of the Classic period, and for this reason escaped the iconoclastic fury of the Conquistadores. It is a place of extraordinary appeal, mostly because of its central part. Here, the Maya priests discovered a huge natural hall with, at its center, a natural column created by the joining of a stalactite and a stalagmite over millennia of water percolation. This natural formation bears an impressive resemblance to a Ceiba, a tree which was sacred to the Mayas. As a consequence, the place became the main shrine of the cave, and people brought offerings in the form of ceramic vessels bearing depictions of the mask of the Rain God, many of which can still be seen (Braswell 2012) (Fig. 1.6).

Similarly, the most important sanctuary of the classical age, the oracle of Apollo at Delphi, is located in an astonishing setting. The terrace of the temple overlooks the entire valley, covered by olive and other trees. To the rear, the temple lies at the foot of Mount Parnassus and is directly overlooked by two spectacular crags, known as the Fedriades, between which a sacred spring spouts from a crack. Ancient sources report that the specific point where the temple was to be constructed was indicated directly by the gods; actually, research has shown the presence under the temple of geological faults producing hydrocarbon vapors which can cause a state of euphoria, that is, the state in which the oracle—incarnated by a priestess, the Pythia—gave its notoriously ambiguous answers (De Boer 2000) (Fig. 1.7).

Fig. 1.6 Balamkanchè, Yucatan, Mexico The "Ceiba tree" formed by natural concretions in the main area of the cave

A further example will be of help in showing the relationship between natural features of the landscape and the cognitive processes which lead to the sacralization of the landscape itself. It comes from Neolithic Europe, precisely from Antequera, Spain, where an enormous Dolmen (stone-chambered tomb) is located. Called Dolmen de Menga, it is was built around the middle of the fourth millennium BC, and is composed by a antechamber, which opens up into a wide chamber roofed with enormous stone slabs and supported by pillars. When visiting the tomb, one cannot fail to notice that the horizon as seen from the tomb is taken up by a rather odd mountain. This mountain, called Peña de los Enamorados, is quite isolated in the landscape and has a peculiar feature: its profile very clearly resembles a gigantic recumbent human head looking up to the sky. This sleeping giant lies only seven kilometres to the north-east of Dolmen de Menga; its connection with the tomb becomes dramatic when the visitor first enters the tomb and then looks back: at this point the human profile fully occupies the view and it becomes obvious that the builders planned the axis of the tomb to obtain this spectacular, permanent hierophany. The Dolmen

Fig. 1.7 Delphi. View of the upper terrace of the temple of Apollo

de Menga was thus deliberately oriented to the mountain, probably as an object of worship (Aguayo and García-Sanjuán 2002) (Fig. 1.8).

A final, but essential, feature of sacred landscapes is their persistence: once a place has become sacred, cultural memory—helped by the interest of power with the input of worldly power—will subsequently confirm the sacredness of that very same place. For instance, if the burial site of a king is recognized as a sacred space, all subsequent rulers will desire to be buried nearby, in order to be associated with their name and memory, and the corresponding landscape will evolve over the centuries, forming a sacred landscape. Usually, specific rules were established to regulate the evolution of these landscapes in accordance with worldview criteria. A key example is what occurred at Giza, Egypt, during the 4th dynasty (between the 26th and the twenty-fivth centuries BC). Here the tombs of the Pharaohs, the pyramids, developed in accordance with two "cosmization" criteria (Fig. 1.9). Firstly, each pyramid had to be oriented with the sides parallel to the cardinal directions (the accuracy obtained by the Egyptians in this orientation is astounding). This assured the correct "cosmization" of the tombs, their correct positioning in relation to the Cosmos, and enabled the

Fig. 1.8 Antequera, Spain. The sleeping giant (Pena del los Enamorados) as seen from the entrance to Dolmen de Menga

entrance corridor—which was always located on the north face—to point towards the circumpolar stars which, as was the case for the Emperors in China, were closely connected with the Pharaoh's power (in Egypt, the deceased Pharaoh was supposed to become one of these stars). Secondly, the pyramids had to be placed in such a way as to form a row along the line of sight from the temple of the Sun in Heliopolis, on the opposite bank of the Nile (see Magli 2013, 2018 and references therein). This led to the—otherwise inexplicable—arrangement on the Giza plateau of the three royal pyramids of Khufu (Cheops) Kafra and Menkaura, whose placement occurred progressively farther in the desert and whose southeastern corners were accurately aligned in such a way as to "merge" their images in a sort of mirage when the sight progressively aligns along the corresponding direction.

Fig. 1.9 Giza, Egypt. A photograph taken from behind the south-east corner of the third pyramid. The axis connecting the corners of the three main pyramids is clearly visible

The Giza Plateau is thus a fascinating example of what a sacred landscape can really be: a place which transmits—at a distance of centuries or even millennia—the feelings, knowledge, and beliefs of its builders, in a vibrant and spectacular manner. The sacred landscapes of ancient China are no exception. As we shall soon see, indeed, beginning with the First Emperor, a series of sites were selected and a series of monuments were built in them, to assure the afterlife to the deceased emperors and to legitimate the power and the rights of their successors. The approach I shall follow here in studying these landscapes is that of Cognitive Archaeology, as developed over the last thirty years or so (Zubrow 1994; Flannery and Marcus 1996; Segal 2009; Rapoport 1982) (Fig. 1.10).

Fig. 1.10 The sacred landscape of Xiaoling, the tomb of the first Ming Emperor. *Credits* Images used under license from Shutterstock.com

In a nutshell, Cognitive Archaeology is an approach focusing on the material relics of the human past, which aims to describe them as objects that had their primary cultural existence as "percepts" in topological relation to one another. They thus fitted the cognitive schemes of their creators, and can be *fully* understood—that is, not only "functionally" understood—only if such schemes are also studied and, at least as far as possible, understood as well. In the case of Chinese sacred landscapes, this cognitive approach turns out to be extremely effective, especially if it is coupled with the use of satellite imagery, as we are going to do in this book. In fact, this combined approach will allow us to establish the mutual relationships between monuments, and at the same time, the evolving dynamics of the constructed features of the landscape, such as intervening buildings, and will also allow us to analyze the "cognitive rules" which were followed to maintain the correct "cosmization".

Chapter 2
In Between Wind and Water

2.1 Form Feng Shui

Feng Shui—literally "Wind and Water "- is a *geomantic* doctrine, that is, a divination practice which surveys the characteristics of a place—morphology, orientation, distribution of waters and vegetation, climate and winds, and so on—to establish whether and to what extent a site under consideration is "auspicious" for the locating of graves, but also for the building of houses and even entire towns. Feng Shui arose from the coalescence of divination and natural philosophy. The period in which this came about is open to dispute, though. Undoubtedly, Feng Shui adopted very ancient symbols and traditions, such as the four directional animals, and, moreover, two of the "fundamentals" of Feng Shui—protecting from the northern winds, having flowing water nearby—look like quite natural conditions for any dwelling, especially on the Loss plateau, so these features may also be extremely old (Eitel 1902; Yoon 2006; Bennet 1978). However, in spite of existing claims, the establishment of Feng Shui as a state-recognized doctrine is a relatively late and slow-moving phenomenon (Bruun 2008). As far as literary texts are concerned, we cannot go back earlier than the chaotic period immediately following the Han Dynasty. In this period lived Guo Pu (276–324 AD), to whom the first known Feng Shui manual, the *Zang Shu* (Book of Burial), is usually credited. This attribution is however uncertain and it may well be that the text is of a later epoch. Indeed, we have only indirect evidence of the existence of geomantic books before the Tang dynasty; during the Tang, we know that Emperor Taizong (626–649 AD) sought to establish which geomantic books could be considered reliable by means of a commission of experts (it appears that the commission was quite skeptical about the whole thing, and even declared some Feng Shui ideas as contrary to the Confucian Classics). In short, what can be asserted with certainty is that Feng Shui developed between the Han and the Song dynasties, being finally established under the latter. This, thus, occurred around the same time as the formulation of Neo-Confucianism.

This movement started within the lively cultural atmosphere created by the Song dynasty, aimed at a revival and updating of Confucianism. The key objective was, on one hand, the rebuttal of Taoism and Buddhism as religions, essentially embracing a rationalist viewpoint, and on the other, the espousal of a series of concepts and values typical of their philosophical frameworks. An influential author in this context was Zhu Xi (1130–1200 AD), whose teachings were adopted, together with the Classics, in the standard background required for national examinations. In Zhu Xi's cosmological view, Qi plays a crucial role, to the extent that all human capacities are considered a byproduct of its flow. The value he attributed to family rituals, in which the longstanding Chinese tradition of ancestor worship was revised and updated, is also particularly important for the development of Feng Shui. Indeed, ancestors' spirits were considered capable of influencing real life, and the belief of the possible intervention in human life of non-human beings—ancestors or gods, ghosts or evils—was profoundly rooted in Chinese thought. However, it is with Neo Confucianism that the cult of the ancestors became a cyclical component of human life: for instance, the *Qingming* (the Feast of the Dead) became a main social event for sacrifices to the dead. The same also holds for Buddhist festivals, which became occasions for visiting graves. Grave rites became in this way official meetings for marking group (clan) identity and lineage, and the importance of the ancestors in turn boosted the magnificence and the desire for the "correct" placement and orientation of their tombs. Feng Shui masters were thus able to build upon the new Neo-Confucian state orthodoxy, and to acquire a stable position by framing their discipline within such orthodoxy. The final version of the doctrine appears as the one formalized in the works by Yang Yunsong (834–900 AD). Simultaneously, however, on account of the invention and improvement of the magnetic compass, a second approach to Feng Shui was also established. To distinguish the two, the first is usually called Form Feng Shui school and the second Compass Feng Shui school, although the two approaches are intertwined and not necessarily mutually exclusive. The Compass approach will be described in the next section, while in the present section we shall pick up the Book of Burial which—notwithstanding the uncertainties already mentioned about its authorship and dating—is a very useful starting point for discussing the basics of the form school (Field 2001; Zhang 2004).

According to the Book of Burial, Feng Shui aims at the identification of the best places for burials by "studying" the flow of Qi in the landscape. The author refers himself to a previously existing "Classic" which might actually be a literary invention; in any case, this Classic says that:

> Burial is contingent upon Qi, and the five (phases of) Qi course through the earth, materialize and give birth to the myriad things.

The idea is that in a place that is optimal for the flow of vital energy, Qi would literally pass along the bones of the deceased, and thereby be transferred to his family, assuring prosperity and health to his living descendants:

> Man receives his body from his parents. If the ancestors' bones acquire Qi, the descendants' bodies are endowed. The Classic says: Qi is moved and responds in kind; the blessings

of ghosts extend to the living…life is accumulated Qi; it solidifies into bone, which alone remains after death. Burial returns Qi to the bones which is the way the living are endowed.

To find such places, Feng Shui relies on the observation of the morphology of mountains and rivers, their orientation and their relationship to winds, since

The Qi of Yin and Yang breathes out as wind, rises up as clouds, descends as rain, and courses underground as vital energy. Earth is the receptacle of Qi—where there is earth, there is Qi. Qi is the mother of water—where there is Qi, there is water. The Classic says: Qi flows where the earth changes shape; the flora and fauna are thereby nourished.

In order to establish favorable places—in line with the main idea that "where Qi is manifested on the surface of the Earth, underground Qi accumulates vitality"—a series of requirements must be verified. First of all, mountains must be covered by vegetation, and must be accompanied by lower hills. Indeed

Bare, severed, bouldery, overreaching, and solitary mountains produce new misfortune and dissolve acquired fortune.

The tallest visible mountain at an auspicious site must be (broadly) to the north, accompanied by chain of hills to the east and west, and by a lower, distant hill to the south. The cardinal directions are identified with the totemic animals of the cardinal points:

Bury with the Cerulean Dragon to the left, the White Tiger to the right, the Vermilion Bird in front, and the Dark Turtle in back…The Dark Turtle hangs its head; the Vermilion Bird hovers in dance; the Cerulean Dragon coils sinuously; the White Tiger crouches down (Fig. 2.1).

So the lateral hills are called the eastern Blue Dragon and the western White Tiger. Their function is fundamental because otherwise Qi "will dissipate in the blowing wind". The specific shape of each hill at a site has to be studied and made parallel to a crouching Dragon and to his body parts. In particular,

Where contour ceases and features soar high, with a stream in front and a hill behind, here hides the head of the dragon (Fig. 2.2).

The role of water courses is also important. In particular, a river should be flowing to the south, since "from a source in the Vermilion Bird vital Qi will spring". Finally, the "cosmic" symbolism of the five elements is associated with the soil:

The soil should be fine and firm, moist and lustrous; it should be cleavable like jade or fat, and composed of all the five colors.

Starting from these main characteristics, a series of further refinements were applied to establish the supposed level of suitability of a site. In particular, the Feng Shui masters developed a complex classification of shapes of hills and types of river curves, so complex, in fact, that one master once wrote that finding a fully-acceptable Feng Shui place is virtually impossible. At any rate, if all (or at least the majority of) such requirements were considered to be met, the place in question was declared auspicious and thus suitable for the building of the tomb, so that ministers, kings, or even emperors in the line of descent of the buried persons could be "produced". Guo Pu says of such places:

Fig. 2.1 An example of a "geomantic chart" showing the characteristics of an auspicious site. From a seventeenth century Feng Shui manual. *Credits* Images in the public domain

Fig. 2.2 An example of a Feng Shui project: Jingling, Qing Eastern Tombs

> Of layers and folds of mountain chains, of ranges of hills and branches of arteries, it is the exceptional that must be selected.

And indeed, *exceptional* is the right word to describe some of the Feng Shui places chosen for the imperial tombs of the Ming and the Qing rulers, as we shall see in the sequel of this book.

2.2   Compass Feng Shui

The attractive power of the lodestone (magnetite) has been known in China at least since the middle of the 1st millennium BC (it was also known in the west, as Aristotle first noted). This knowledge does not, however, in itself imply awareness of the "directive" property, namely the fact that a piece of lodestone, or a piece of iron magnetized by contact with lodestone, when left free of "floating", will orient itself along a rough north-south direction (for a discussion of the Earth's magnetism see the Appendix). The discovery of the directional property clearly led to the implementation of some form of orientation compass, of which the familiar one is based on a magnetized iron needle floating around a small support. This object did not appear in the west until the twelfth century BC, while it had already been known in China for some centuries, as we shall shortly see. However, a "magnetic compass" does not necessarily have to be a free floating magnetic needle, and there is evidence that *another* form of magnetic compass was invented in China long before, under the Han dynasty to be precise.

Some confusion about this problem stemmed from an object called in ancient Chinese texts "south-pointing carriage": a pointer mounted on a cart and able to maintain an originally fixed direction (say, south). Clearly, it is natural to suppose that it was a compass-based mechanism; however Needham (1965, 1970) has convincingly shown that this carriage was *not* based on magnetism, but involved mainly a system of gears and wheels so that the pointer maintained its direction by continually compensating for any diversion of the vehicle away from that direction (nonetheless, personally I do not believe that this object could really work without being subjected to any adjustment of direction from time to time, by using astronomical orientation or indeed using a form of magnetic compass). At the same time, Needham himself came up with a quite unexpected and intriguing picture of how ancient Chinese compasses might have looked like. This picture is based on a couple of texts of the Han period and on a very curious stone relief. The main text in this context was written by Wang Chong (27–100 AD). Wang was a scientist and a philosopher famous for his skeptical and rational attitude towards the laws of nature. In his *Lunheng* ("Critical Essays"), he confutes superstitions and divination, criticizing Taoism and adopting a viewpoint that in modern terms might be termed agnostic. His works are very important and in some ways pioneering as regards the development of the scientific method. They are, however, mentioned here because they appear to be extremely significant not only for the history of science in China, but also for the history of something that Wang

Chong certainly would have deemed to be based on superstition, that is, geomancy. In the passage of the Lunheng which interests us, Wang Chong strongly criticizes the fable that there exists a plant able to orient itself to people in the vicinity. He writes:

> As for the indicator-plant, it is probable that there was never any such thing, and that it is just a fable. Or even if there was such a herb, it was probably only a fable that it had the faculty of pointing at people...but, when the south-controlling spoon is thrown upon the ground, it comes to rest pointing at the south.

In this passage, to explain that the purported plant does not exist, he contrasts it with a similar thing which does, in fact, exist. It is a spoon which has the ability, if moved, to come to rest pointing south. This text is of particular interest, especially since Wang Chong is a very scholarly, erudite author. It is for this reason that Needham has taken it very seriously, proposing that the spoon was the first version ever developed of a magnetic compass. As a matter of fact, Chinese pottery spoons of the Han dynasty period are well-known archaeologically and are really much more similar to ladles, whose end concave part is quite thick. It is, therefore, conceivable that the Chinese could have carved a spoon out of lodestone (natural magnetite) in such a way that the weight of the end part kept the object in equilibrium on a flat surface with the handle upright and the point of contact with the surface reduced to a minimum. A spoon carved in this way, when laid delicately on a flat horizontal plane, will orient itself along the magnetic north-south direction. The similarity between a ladle and the Great Bear, and the fact that the Chinese actually used this constellation as a sort of clock-hand during the night, led Needham to hypothesize that the surface on which this first form of compass was put was precisely the center of the Shihi, the diviner's board, where—as we have seen—the Great Bear was always represented. Thus, the magnetic spoon compass was not used for practical means (e.g., navigation) but—if it existed, as it is probable—was part of the equipment of diviners and magicians. Experiments have actually been performed with lodestone extracted from a source available in Han times, and it has been shown that this "Needham compass" works in a reasonable way, since the magnetic torque is able to overcome the small friction between the bowl and the underlying smooth surface. It is fundamental, however, to have a very flat support and to carve the contact region accurately.

Needham's "magnetic spoon compass" has been generally accepted as a true fact and reproductions of the object are ubiquitous. I should say that I am not 100% convinced, as not even one such lodestone spoon has been recovered, while the object would seem to be a perfectly natural one to add to funerary equipment (and indeed, examples of diviner's boards are known from tombs, as mentioned). However, I should also say that I do share Needham's astonishment when he discovered the existence of a Han-epoch relief which seems to represent precisely this object (Fig. 2.3).

It is a stone relief conserved in the Museum Rietberg, Zurich, and dated to around 14 AD. It depicts an exhibition (either sacred, or profane) which is taking place at the emperor's court. The central scene shows people dancing around a central drum (according to Needham, the figures may actually be mechanical dummies on a roundabout or carousel). Seated figures, probably the royal court, are represented

Fig. 2.3 Stone relief probably representing court's entertainments, around 20 AD, Han dynasty. *Credits* Images Courtesy Museum Rietberg, Zurich, used under kind permission

on the top row. To the right, a person draws attention with his hand to a kneeling figure who is looking at a spoon resting on the flat surface of a small square table. To the left, a formally similar scene seems to be the exhibition of some kind of doll or automaton. It is frankly difficult to believe that the sculptor would use a entire register of his work of art to represent an attendant preparing or bringing drinks or food with a spoon, and it is much more natural to think that the scene represents one of the "wonders" presented to the Emperor, which might well have been a lodestone compass spoon (Fig. 2.4).

To complete the evidence pointing to an early version of the magnetic compass, I mention the curious description, present in the Han annals and again noted by Needham, of the last minutes of life of the usurper Wang Mang—a Han Dynasty official who seized the throne and attempted to found his own dynasty—before his capture in 23 AD. The imperial palace has almost been taken over by enemies, when he

> …held in his hand the spoon-headed dagger of the Emperor Shun. An astrological official placed a diviner's board in front of him, adjusting it to correspond with the day and hour. He turned his seat, following the handle of the ladle, and so sat. Then he said, 'Heaven has given the empire to me; how can the Han armies take it away?

The reference to Emperor Shun is a direct hint at the mythical history of ancient China, as has been handed down in texts from the Warring States period and

Fig. 2.4 Stone relief probably representing court's entertainments, around 20 AD, Han dynasty. Detail of the upper register: a kneeling man is showing what seems to be a spoon magnetic compass. *Credits* Images Courtesy Museum Rietberg, Zurich, used under kind permission

reported by Sima Qian. According to this myth, the first rulers of China were "Three Sovereigns and Five Emperors". The first of the five ancestral emperors is Huangdi, the so-called Yellow Emperor, whose life is placed at the beginning of the Shiji history book. Several inventions were attributed to the Yellow Emperor, from musical instruments to writing and the calendar (his reign is traditionally thought to have been in the twenty-seventh century BC). The name Huangdi was adopted by the First Emperor of Qin, and his cult further developed during the Han, who were searching for an ancestral icon of centralized power. According to Sima, the Yellow Emperor had four successors, and the usurper Wang Mang apparently claimed to be a direct heir of the last one, Shun, to whom the object he holds in the hands as a sign of power originally belonged. It should be noticed that in this text the "ladle" cited immediately afterwards need not necessarily be connected with the "spoon" and might be simply taken as a synonym of the Great Bear, which we do know was represented on the diviner's board. However, the fundamental direction associated with empire was south—in the sense that the emperor was supposed to be "the north" and to guard the empire towards the south—and so this should be the direction indicated by the ladle. All things considered, then, this scene (textual in this case) is also perhaps alluding to a Needham spoon compass placed on the diviner's board.

Leaving aside the precise dating of the discovery of the magnetic compass, we now need to examine a point which—I do not know why—is usually neglected or taken for granted in most the literature, namely what happened when the directional property was discovered. This occurred in a period and in an environment in which the court intellectuals and philosophers were debating and constructing the complex scenario of natural philosophy, religion and science which we outlined in the first sections, and we can extract two fundamental facts from it:

- Qi was supposed to flow everywhere
- The main cosmographic direction that informed all Chinese thought and, above all, was closely linked with the Emperor's image, was north-south or, more precisely, north-to-south.

Thus, being of course unaware of the physical explanation, the discoverers of the magnetic compass *must have thought that they had discovered an instrument able to measure the flow of Qi*. It must have been a dramatic discovery, and it is therefore natural that such a groundbreaking instrument was merged into divination, as suggested by Needham. Later, this was easily accommodated into Feng Shui, because the direction indicated by the compass matched the prominent role of the north-to-south direction in Form Feng Shui (Chinese compasses are always made to point south). When this really occurred is again a matter of debate: Guo Pu does not refer to magnetic orientation and indeed the first clear evidence of a Compass Feng Shui tradition are more or less contemporary with the invention of the *magnetic needle* compass. This occurred when someone noticed that a handy, simple instrument could be constructed using a piece of iron (which must originally have had the shape of a fish) which had been magnetized by being rubbed on magnetite. This invention also appears to fit into a geomantic context. Indeed, although the first text describing the use of magnetic needle compass for orientation is from the eleventh century BC, the magnetic needle compass (called "mysterious needle") had already appeared around 900 AD in a geomantic text by Ma Gao (interestingly, in the eleventh century, the Chinese had also a clear conception of the fact that the magnetic north-south is different from the astronomical one, and all this was long before the first European mention of the compass).

To sum up, it is reasonable to conclude that (Needham 1970):

The magnetic needle was in use for determining compass directions, no doubt mainly for geomancy, by the middle of the tenth century, and in all probability some centuries earlier.

Compass Feng Shui developed the magnetic compass in a form—known as *Luopan*—which can really be called "geomantic compass". I am one of the fortunate possessors of a Luopan, which was (a bit ironically) gifted to me by my Chinese students of Archaeoastronomy in Xi'an, and I can state that it really is a very complicated object (Fig. 2.5).

In the center, there is just a standard, south-pointing needle. However, the object is provided with a series of embedded concentric rings bearing symbols and words (there can be up to 40 circles). Things are further complicated by the fact that there exist different kinds of Luopan (that is, different sets of circles) and that each Feng Shui master may design his own version. The circles are divided into several different sectors (number of sectors can vary as well) in such a way that, once the south direction is known, one can read information on every desired direction. The reason for doing this is that, as we have seen, in Feng Shui practice all the directions characterizing a site—flowing of rivers, orientation of hills' veins, and so on—are considered. In Compass Feng Shui, therefore, the same directions are compass-measured and to each of them a degree of auspiciousness is assigned, depending on the sector(s) of the Luopan to which they happen to belong (Needham 1956, 1965; Campbell 2001; Yi 1994).

Fig. 2.5 A modern example of a Luopan, the geomantic compass

2.3   The Establishment of Feng Shui

The Song period came to an abrupt end with the Mongol invasion and the establishment of the Mongol rule under the Yuan dynasty (1279–1368 AD). However, it
appears that geomancy and other Chinese divining techniques were easily adopted
by the invaders, who merged their own beliefs with the Chinese Taoist and Buddhist
syncretism. As a consequence, Feng Shui survived well up to advent of the Ming
Dynasty. Under the Ming, we see a period of intense flourishing of all divination
practices. The beliefs of the ruling class were syncretist to the greatest degree; as
well as advocating the "three teachings" (Confucianism, Taoism and Buddhism),
they openly believed in ghosts and spirits, and placed special faith in divination as a
means of receiving spiritual guidance (Smith 2012).

Encouraged as it was at the highest level, Feng Shui spread into lower social
classes, and was used for identifying auspicious places for dwellings and for tombs
which mimicked those intended, for instance, for royal burials. However, the "elastic"
nature of the contents and the foundations of the doctrine remained and several
masters continued to propose modifications and personal methods over the course of
the centuries. The two schools also assumed geographical characters, as people living
on flatland, of course, favored Compass Feng Shui. The debate between schools and
masters led to attempts at establishing "best practices" and to the publication of a

series of manuals and guides. In this respect, a curious and interesting book, probably written in the early Ming period, is the text *Twenty-four Difficult Problems* (Paton 2013). The book appears to be organized didactically in the question-answer format and—just to be clear—begins (Problem 1) citing its own "Classic", which for the author is none other than the Book of Burial, and reasserting that

> In an area of mountains and rivers there must be a mountain which is the tallest and most exalted and which is dominant. This is called the ancestral mountain. In terms of the direction of this mountain, even though there are branches from each of the eight directions, more are certain to emerge at the face of the major configurational force; the mountain subsides and rises again, breaks off and reconnects, but is certain to face water at the front and follow it so that they hasten forward together. Where Qi focuses, the form must swerve and there is interaction with the water; all others rush back along with the configurational force of the mountain... The rocks are the bones of the mountain.

Some of the other "difficult problems" are really technical...for instance, the problem of understanding if and when a flow of water is interrupting a vein of Qi coming from a mountain. In general, however, all the answers given are based on forms, and indeed in Problem 8, the author—clearly a Form Feng Shui master—dispatches the compass approach, saying (about the compass bearings) that:

> The sages established a means to teach men to distinguish position so that the populace would not lose their bearings and that is all. Auspiciousness and inauspiciousness have never been a part of it.

Interestingly, the author also ventures to make observations about the flow of Qi on a more extensive, global-geographical scale, showing that in those times Feng Shui was upgrading itself to the level of "Geomantic Geography". For instance, he relates the bends and the morphology of the flow of the Yangtze with the wealth and importance of the cities along its course.

Finally, and with a certain emotion, we can read (Problem 5) how a geomancer is supposed to work, and imagine the court chief geomancer when he first climbed the Tianshou mountain and saw the beautiful valley which was to become the Ming Royal Necropolis (Sect. 7.1) as well as, six centuries later, the UNESCO site of the Valley of the Thirteen Tombs. Indeed:

> Problem 5: Which are the most important methods of seeking the dragon, observing the configurational force and examining nodes?
>
> Reply: First look to an especially high place that the ancestors esteemed. Next, examine which branch amongst the multitude first breaks. Where qi is cramped and clustered as it emerges is the place of the main dragon. Where the dragon comes to the end is a node. There must be much congealing of movement and branching. Even if a dragon vein is not long, it will control a luxuriant and thriving cover. The trees that cover the qi of a dragon are essentially noble like kings and lords. Their offspring will not lose their peace, wealth, honor and glory. This is what is meant when practitioners of the art say that to build a family it is necessary to have a good husband and wife.

Fig. 2.6 The impressive scenery of Tailing, Qing Western Tombs. *Credits* Images used under license from Shutterstock.com

Fig. 2.7 A Feng Shui project of about 1880: The tomb of Empress Dowager Cixi, Qing Eastern Tombs

The fall of the Ming saw the advent of another foreign Dynasty after the Mongol one, the Manchu Qing. The Qing adopted Chinese culture and beliefs and there is clear evidence that Feng Shui became even more potent and pervasive under them. For example, Feng Shui was discussed in depth in the Encyclopedia produced in 1726 and was also repeatedly noted in accounts by Western writers, who, biased as they could sometimes be, certainly noticed a widespread and important phenomenon (Figs. 2.6 and 2.7).

Last but not least, as we shall see, during the reign of the Qing Emperor Kangxi, the Feng Shui "Geomantic Geography" was endorsed as a means of consolidating the Manchu Mandate of Heaven, claiming that the Qi of the country was generated in the Manchu sacred mountains. Feng Shui maintained its important role up to the end of the Imperial period, as a visit to the tomb of Empress Dowager Cixi (who died in 1908) in the Eastern Qing Necropolis (Sect. 8.2) can easily verify, and has actually survived up to present times (Bruun 2008; Coggins 2014; Coggins et al. 2012, 2019; Miller 2016).

Chapter 3
A Mound and A Terracotta Army

3.1 The First Emperor and the Mandate of Heaven

After a long war, with the final battle of 221 BC, Ying Zheng, the king of the state of Qin, proclaimed himself *Shihuang*, the first Emperor, and became the ruler of a territory twice as large as modern France and Spain put together (in this book, historical information on the Chinese emperors is mostly based on Paludan 1998). This territory was further extended by his generals over the years. Unification of the state did not mean unification of people, though, and Shihuang made tremendous efforts—some of which were quite brutal—in this direction.

First of all, economic and political reforms were put into place to standardize measures and procedures, making the state bureaucracy a very efficient mechanism. The empire was divided into "commanderies", and units of measurements were unified, as well as the cart gauges. The country was criss-crossed by a new network of roads and channels, and currencies and scripts were unified as well. Huge state projects were devised: among them, a boundary wall—mostly made of rammed earth—which may be considered as a sort of early version of the so-called Great Wall (not to be confused with the Ming one in bricks and stones which was constructed 1700 years later) to protect the northern borders. From the intellectual point of view, however, the first Emperor's reign has come down to us as appearing a period of obscurantism. Shihuang rejected the subtleties of Confucianism, adopting a rather brutal legalistic system (obey or be punished). At least according to Sima Qian, a number of classical books was banned and burned, and hundreds of scholars buried alive, although this last fact might be a legend. Historical accounts further report that in the last years of his life Shihuang became obsessed by death and started a search for an elixir of life which—needless to say—no one was able to provide. Apparently, he became convinced about the validity of legends that identified imaginary islands in the east as places of immortality; he personally made three visits to Zhifu Island, an apparently insignificant islet in Shandong Province, associated with legends and perhaps considered as a good starting point for further exploration in search of the

© The Editor(s) (if applicable) and The Author(s), under exclusive license
to Springer Nature Switzerland AG 2020
G. Magli, *Sacred Landscapes of Imperial China*,
https://doi.org/10.1007/978-3-030-49324-0_3

elixir of immortality (the people sent on expeditions to search for it judiciously decided not to come back to report to the Emperor). It was perhaps this lack of answers that lay at the origin of his animosity towards intellectuals; anyway, in a sense, it was Shihang's rough approach to normalization and his insane grandeur that prepared the way for the rapid fall of the Qin dynasty (occurring with his son) and to the splendor of the following one, the Han dynasty.

Shihuang's efforts were of course also devoted to the symbolic foundation of his power. To this end, he revived and strengthened an old cornerstone of the Chinese ruler's power. A seal carved on the occasion of Shihuang's accession, indeed states that the first Emperor has received the *Tianming*, the Mandate of Heaven. Numerous cultures in the world have accepted the power of their rulers as being bestowed by the Gods, and China is no exception. Since very ancient times, Chinese rulers justified themselves as having been chosen by *Tian*—the deified sky, identified with the celestial order and the regularity of the celestial cycles—to be the keeper of the very same order on earth. The content of this *Mandate of Heaven* was subjected to different interpretations over the millennia; it started as the attribution of the role of chief priest to the king, but evolved to being the identification of the ruler with the divine manifestation of the deified sky itself (Sivin 1995). As a consequence, extremely complex and profound connections appear to exist between key astronomical events and key turning points in pre-imperial Chinese history (Pankenier 1995, 2013). This book is not the place to go into details of such issues; one very important thing, however, should be noted. Contrary to the situation with many European monarchies, where the "mandate" was considered as bestowed on a ruling family unconditionally, for once and for all (the so-called Divine Right of the Kings), the Chinese mandate could be withdrawn. In practice, this means that overthrowing of a ruler—due to rebellions, wars, or political instability due to famines or droughts—was interpreted a posteriori as an indication of his having lost Heaven's favor. Thus, the mandate could pass to other rulers, not necessarily of noble origin (this occurred, for instance, at the start of the Ming dynasty). The mandate was, therefore, to be held with the utmost care, and implied a series of obligations, such as ceremonies, processions and other rituals, besides the duty (not always adhered to) of being a just monarch.

In the case of the first Emperor, it appears that—at least in his own mind—the Mandate of Heaven was in some sense also to be conserved in the afterworld. This idea was to lead him to conceive and build the first of the sacred landscapes of Imperial China: that of his own mausoleum, located in Lintong, west of the Qin capital Xianjang and south of the River Wei. Taken as a whole, this mausoleum is such a huge, complex structure that a complete description would require an entire book, so that we shall restrict ourselves here mostly to the cognitive aspects.

Very generally speaking, the complex was conceived of as a symbolic replica of the imperial town, with above-ground inner and outer enclosures, and a series of underground pits and galleries containing replicas of the living world familiar to the Emperor. Walls were built with rammed earth, so that only scant traces remain. However, what can be seen is enough to show that the funerary town was really huge: a rectangular enclosure around 2200 × 1000 m, with an inner rectangular wall around 1300 × 600 m. The tomb itself is located under an enormous square-based

Fig. 3.1 Lintong, the huge mass of the Shihuang Mausoleum, covered by trees

mound (350 m of side base, more than 50 m high) which occupies the southern part of the inner precinct (Fig. 3.1).

The interior of the tomb is unexcavated, but many pits and galleries have been discovered and brought to light in the surrounding area; perhaps others await discovery as well. In any case, what has been unearthed is amply sufficient to glimpse the incredible complexity—both practical and symbolical—of the funerary project of Shihuang. Of course, we shall start our visit from the most famous of these discoveries.

3.2 The Terracotta "Warriors"

For some 2000 years, the mausoleum was left undisturbed, and farmers had worked in the surrounding area apparently without finding anything relevant. Yet in March 1974, at a distance of 3.5 km from the mound, people digging a well encountered a stratus of terracotta shards and bricks, with sparse bronze arrowheads. In the space of a few months, it became clear that what they had found was the result of the collapse of a gallery originally containing many rows of terracotta statues of men. Following the standard terminology, we shall call them warriors, although this definition will need to be revised later. Today this place is called Pit 1; three other pits were discovered

Fig. 3.2 Lintong, Mausoleum of Shihuang. Terracotta warriors, Pit 1, front view

nearby in 1976 (of these, Pit 4 is empty). Altogether, more than 7000 terracotta warriors have been uncovered in the galleries, as well as the remains of about a hundred wooden chariots, each one with its (terracotta) horses (Sun 2002; Xu 2015; Xiuzhen et al. 2016; Wei and Weixing 2015) (Figs. 3.2 and 3.3).

These statues have never been mentioned in ancient Chinese literature; in particular, they are not mentioned in the *Records of the Grand Historian,* the general history of China written by Sima Qian (Watson 1993). So, we cannot rely on written sources for our understanding of this astonishing archaeological find. Memory of the statues was not lost immediately after Shihuang's death, however, as there is no doubt that the galleries were sacked and set on fire during the rebellion which caused the fall of the Emperor's son and the rapid end of the Qin dynasty.

A visit to the warrior's Pits is an unforgettable experience. The galleries are literally crowded with statues; the floors are paved and the ceilings are supported by huge wooden structures covered by reed mats. In Pit 1, 362 × 62 m wide but divided into parallel galleries by walls of rammed earth, the warriors stand in rigorous order, all facing true east, except for an external line where soldiers face outwards on any side. In Pit 2, measuring 124 × 98 m (mostly still to be excavated) there is what seems to be a carted cavalry unit with an estimate of around 90 chariots and corresponding horses, together with a company of archers and other troops.

Finally, Pit 3 is much smaller (29 × 25 m) and contains something that looks like a command unit. It actually hosts a unique (dissolved) wooden chariot with four terracotta horses, and 68 warriors. It is very clear that the owner of the chariot is not present, and that all people in the pit are paraded to wait for him (Fig. 3.4).

Fig. 3.3 Lintong, Mausoleum of Shihuang. Terracotta warriors, Pit 1, side view

The role of each statue (soldier, officer, archer, general) can be recognized from pose, dress and hairstyle. Studies have shown that the organization of this "ghost" army corresponds, at least roughly, to what we know of the organization of a Chinese army unit in Qin period. The warriors are today unarmed, but it appears that they were carrying bronze weapons and that most of these weapons were stolen in rebels' incursions, since a good number of scattered bronze swords and arrowheads has been found in the pits. The statues are all life-size or—perhaps—slightly more so, since their average height is about 178 cm (Komlos 2003). They are not portraits of specific people, though, as they are assembled from separate parts which include the feet, the legs, the torso, the head and so on. Each part was created in several different ways, and the face details of the statues were modeled in clay, generating a great number of possible combinations. All the warriors were meticulously painted, although in most cases today only traces of pigments remain. The statues are an unicum in Chinese Art history: very few and small terracotta statues are known from the period before, and life-size models would never be replicated in terracotta again. This has provoked considerable academic debate about the possibility that life-size statuary was imported from the west: the Hellenistic world, if not directly Greece. There is no serious proof of this hypothesis, however, and I personally tend to see the Qin funerary project as one of those examples in which human genius—united with virtually infinite economic resources—reveals itself in a complete break with

Fig. 3.4   Lintong, Mausoleum of Shihuang. Terracotta warriors, Pit 3

what already exists (another example is what happened in Egypt around 2600 BC with the funerary project of Cheops' father Snefru, see Magli 2012).

This, in short, is the phenomenon known as the terracotta army. But what exactly is the terracotta army, and why was it devised? A very standard interpretation, reported almost everywhere, identifies the warriors as a magical defensive force, intended to act in the afterworld against the enemies of the emperor (or rather, the ghosts of the enemies). The Qin's cruelest battles of conquest were directed eastwards, and this may be the reason why the warriors are only on the eastern side of the mausoleum's mound and tomb, and look towards the east.

In my opinion, although there is no doubt that the army *is* a magical device, it is very doubtful that it is a defensive one; so let us analyze separately the two points. First of all, it should be be noted that modern views usually tend to distort ancient ideas of magic or, to put it in another way, our idea of magic is usually quite different from that of most of the ancient civilizations. For instance, for an ancient Egyptian, nothing would have been more absurd than seeing a movie with a mummy brought back to life and walking, and for an ancient Chinese nothing would be crazier than seeing a walking terracotta warrior. Communication between the two worlds, though it existed and was, indeed, fundamental, was mediated by magical objects which were purely symbolic—a classical example being the false doors in Egyptian tombs (false doors and windows will also appear in Chinese tombs, as we shall see). In Chinese

beliefs, spirits (or rather, what *we* would call spirits) were made up of breaths, and—
in a sense—death was a state of transition, to the point that a much-quoted statement
by Confucius recommends caution: better not to treat the deceased as entirely dead,
although it would not be intelligent to treat them as alive. To help us understand the
Chinese afterworld, we might refer to an extraordinary depiction of it coming from the
funerary equipment of the Mawangdui tombs already mentioned in Sect. 1.2. Dated
to the early Han period, these are the tombs of the Marquis of Dai, his wife and his son
(Buck 1975; Wang 2011; Guolong 2005). In tombs 1 and 3, on top of the (innermost)
coffins, silk funerary banners were found. These are large (approximately 2 mt ×
90 cms), T-shaped garments decorated with elaborate multicolored paintings, which
were probably displayed in the funerary procession. The iconography represented
in the paintings is very complex, but conveys well the ideas underlying the Chinese
conceptions of afterlife. The most famous and elaborate is that of Lady Dai. Here,
the bottom register depicts the underworld: unfriendly creatures like snout-nosed
fish, snakes and antlered animals dwell in this region. Above the underworld lies
a sort of terrace—the human realm—where some figures, probably relatives of the
deceased, are performing a ritual. Above that, another platform shows a central
figure—Lady Dai or perhaps her summoned spirit—attended by five females. The
person represented is unquestionably in the Afterworld, but is depicted as she was in
life, an old lady using a walking stick, and, as matter of fact, her walking stick has been
found in the tomb's equipment. Finally, above this scene, a complex representation
of Heavens is depicted. A bird with human face is shown flying along the upper
reaches of the sky, while two large, intertwining dragons extend along the sides.
Two guardians wait at the gates of the upper, heavenly realm. Inside it, to the left,
a crescent moon is shown with two animals, a hare with pupil-less eyes and a toad,
both associated with the moon in Chinese myths. Below is probably the goddess of
the moon. To the right is the red sun inhabited by a mythological, crow-like bird. At
the very center sits an imposing personage, recognizable as 'the Grand One', a god
responsible for the cosmic cycles (Fig. 3.5).

Given the transitional status of the deceased, the objects buried with the dead did
not necessarily all have to functional: for example, vessels were not suitable for actual
use, and similarly—as we shall see shortly—the chariots of Shihuang were half-sized.
So, yes, the warriors too did need not to be real warriors and yet they appear entirely
appropriate for their magical function. But, what was this function? Their expression
is solemn, their posture static, their gaze almost lost. They do *not* look—at least to
yours truly, the author—like first-line soldiers focusing their attention on the arrival
of enemies, and indeed, on a more detailed scrutiny, they are *not* ready for a battle at
all. Qin soldiers would be wearing solid helmets rather than elaborate and differing
hairstyles, and iron as well as steel (rather than bronze) weapons were at the disposal,
at least, of the elite of the army. What is more, real battle formations certainly did not
had weak external lines guarding the flanks and arranged like bodyguards protecting
music stars at a concert. Finally, it seems that the terracotta warriors could possibly
be equipped for fighting spirits properly—that is, with stone, which repels spirits
in Chinese traditions—but only if necessary. Indeed, in 1998, in the area near the
Mausoleum precinct (today called Pit K9801), an astonishing discovery was made. It

Fig. 3.5 The Silk Banner of Lady Dai (drawing). The details are very complex but the underworld (below). the transitional zone—with the deceased with her stick—in the middle, and the Heavens guarded by two ascending Dragons on top are clearly recognizable. *Credits* Images in the public domain

is a huge pit (130 × 100 m), in which a great number of small stones with tiny holes inside was found. It was then realized that the stones were nothing but delicate little pieces of stone war equipment: armors, helmets, and also horses' armored bridle reins. These pieces were once perfectly fitted together with copper wires and the objects they formed hung on dissolved wooden stands. One of the suits of armor has been painstakingly reconstructed, and it turned out to be made up of around 600 little stones. It has been calculated that a highly skilled stone artisan (only such an artisan could cut so many different kinds of stone pieces, some of which of are millimetric in size, and drill in them the necessary holes for the wires) would take up to one year, working eight hours a day, to produce just one armor (Lin 2007). On the other hand, one only has to see the reconstructed specimen to conclude immediately that these are purely symbolic armors, as only a person of unsound mind would dream of wearing such a cumbersome and heavy (around 25 Kilos) object. So again, this appears to be magical equipment, to be worn in case of need. By who? Well, excavations are still underway, but it may well be that the stone suits of armor amount to thousands, just like the warriors… (Fig. 3.6).

But then, if the warriors are not supposed to fight, what exactly is their function? The key probably lies in the fact that, although the statues are not waiting for enemies, they *are* waiting: they are waiting for the person who owns the main chariot located in Pit 3. This person is supposed to conduct a ritual, perhaps a parade with a crowd of people, and this is why—I surmise—the flanks of the army are guarded. At this point, the conclusion is unavoidable: the warriors are guards and members of the imperial court, and the person they are waiting for is none other than the Emperor himself (Liu 2014, 2015). Associated with this interpretation, a long-standing question exists, which regards sacrifices of living beings on the occasion of royal funerals. Indeed, there is no doubt that there existed in China a tradition of sacrificing horses and humans—concubines and court officials—on kings' deaths. This tradition is reflected

Fig. 3.6 Lintong, Mausoleum of Shihuang, reconstructed stone armour

in archaeological findings of burials of kings of the Qin and of other states before unification. It is difficult to say to what extent this practice was applied, or how much of an honor (and a free pass for afterlife) it was to be buried along with the king. In any case, although sources report that the practice of sacrifice was banned in Qin state in 384 BC, there is no doubt that it was performed again on the occasion of the first Emperor's death. First of all, evidence of animal sacrifices has been found at the Mausoleum's site. These sacrifices are located in pits found in Jiao Village, east of the mausoleum. They contain (don't ask) both terracotta horses and the remains of real horses, perhaps buried alive. A further pit near the inner west gate contained hundreds of equine bones guarded by terracotta grooms (don't ask). Other pits again contain the bones of rare or hunting animals, such as deer, carefully buried in specially prepared coffins (again, don't ask). Sadly, pits containing tens of skeletons of women (almost certainly sacrificed concubines) have also been found, as well as the grave pits of workers, perhaps killed to keep the secret about the tomb's interior. Yet court and military officers have seldom been found. So again, the terracotta warriors (to which, as the writing of this chapter proceeded, I have been more and more tempted to append quotation marks—to the point where I actually did in the title) might really be magical substitutes for the thousands of sacrificed court attendants the burial of such an important personage would have required. Further research might possibly lead to an understanding as to which official occasion they are supposed to be attending. At any rate, the idea that the warriors are waiting for the Emperor is corroborated by other important finds made in two other pits.

The first is the pit closest to the mound that has ever been discovered, located on the west side. It was found to contain two bronze chariots, richly decorated and painted, each complete with four horses and a charioteer (Swart and Till 1984) (Fig. 3.7).

Actually, these are models of chariots, as their dimensions appear to be half life-size. In accordance with Chinese views about afterlife (besides being cast in bronze), they could not be used, and yet, they are really perfect, with all details faithfully copied. The two chariots appear to form a procession which proceeds westward from the burial mound. Chariot 1 opens the procession as a sort of vanguard. It is an open air vehicle, so the charioteer of course needs a large parasol to shade himself from the sun (the umbrella is flexible and can be adjusted to different angles). The object in its entirety (difficult to define it as a bronze sculpture, as it is made of thousands of pieces including bolts and horse's reins) is more than 2 m long. Chariot 2 is a closed cabin vehicle, with horses adorned in gold and silver. The cabin has three rectangular windows, one of which enables the passenger to give orders to the charioteers, while the other two are at the sides. The cabin is accessed from a door at the back, which of course can be opened and closed (like the windows, for what matters); the carriage features elegant painted decorations of stylized clouds. Both the drivers are very clearly ready to go upon a simple order, but the cabin is empty. So again here, we find a symbolic representation of people waiting for the emperor, this time on the way to the west, perhaps towards homeland.

The second important finding is Pit K0006, again close to the mound, in this case near the south-western corner. Usually called the pit of the civil officers, it contains a ramp leading to a (now vanished) wooden chariot, ready for use with four

Fig. 3.7 Lintong, Mausoleum of Shihuang. Empty cabin bronze chariot with bronze driver ready to go

(skeletons of) horses, and 12 life-size terracotta statues. Besides the charioteers, the other statues can be identified as high-ranking officials or magistrates. In particular, one of the figures is holding a ceremonial ax, and some others seem to have been holding between the left arm and the body a (dissolved) bamboo slip. These slips were used to record official data such as taxes or censuses. At the back of the same pit, dozens of horses were buried. Again, the whole scene gives the impression of people waiting for the owner of the chariot, in order to begin an official duty.

Official duties were thus clearly envisaged for the emperor's afterlife, with specialized terracotta statues in attendance. Proceeding with our visit to this gargantuan dream of immortality, we shall now encounter two pits that are—at least at first glance—related, rather than to official duties, to more mundane activities (Duan 2007). The first (labeled K9901) is in the area near the south-eastern corner of the outer wall. It is T-shaped and contained eleven color-painted terracotta statues and a bronze *Ding*, a huge, heavy (more than 200 kilos) cauldron tripod used in ceremonial occasions. Most of the statues of this pit are usually defined as "acrobats" and indeed, they have nothing to do with the "warriors". They look as though they are involved in dances or acrobatic exercises, although their equipment (perhaps ropes, balls, and/or wooden poles) has now vanished, so that understanding what they are really doing is difficult. There are statues, however, depicting other characters: one, in particular, certainly represents a giant person, perhaps a weight-lifter, and, indeed, in the pit, lead and stone weights have also been found (Fig. 3.8).

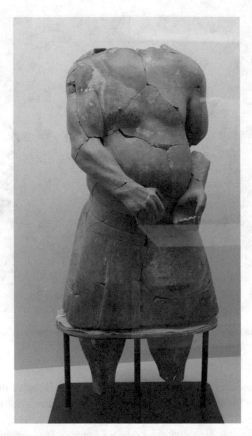

Fig. 3.8 Lintong, Mausoleum of Shihuang, Terracotta "weightlifter"

If K9901 looks like a sort of circus troupe, Pit K0007—found in 2000 near the northeastern corner of the Mausoleum—was also perhaps a section devoted to the emperors' entertainment. The pit occupies a total area of about 978 sq m, and is shaped like the capital letter "F". When they unearthed it, the archaeologists could not believe their eyes: the scene that appeared before them after 2200 years was that of a river bank, or small lake, populated by 46 life-sized bronze birds—cranes, swans, and wild geese—of exquisite craftsmanship (if you look carefully, you can see that one of them has just caught a bronze worm in his beak). In addition, 15 terracotta statues were found. The figures appear to relate to some activities whose implements or equipment, however, have not been conserved. They might be musicians—but the wooden instruments are lost—or hunters of living birds (catching them through rope nets), or both (Fig. 3.9).

As mentioned, a first interpretation of the contents of these two pits might simply be that these are entertainers at the emperor's court, transferred into the afterlife.

Fig. 3.9 Lintong, Mausoleum of Shihuang, bronze bird

However, yet again, the fusion of different elements and of different materials—bronze birds with terracotta attendants, bronze cauldron with terracotta lifters—hints at a magical dimension which most likely goes beyond mere appearance. In this respect, an intriguing and convincing hypothesis has been formulated. The idea is that the "acrobats" of pit K9901 might in reality be engaged in some sort of ritual representation in which the key element is the Ding, the giant cauldron. Sources, indeed, report that Shihuang ordered a thorough search for a lost ancient cauldron of the Zhou dynasty, apparently considered as a mythical, powerful object. This object was connected with the advent of Autumn, and rituals based on Dings for this change in the seasonal cycle are testified to during the Han period (Bulling 1966). Perhaps the other "entertainment" pit, in fact, also had ritual significance, connected in this case with the arrival of the Spring, heralded by migratory birds. This interpretation would tend to contextualize these two "anomalous" pits in the general symbolic significance of the mausoleum microcosm: rites connected with seasonality were included among the duties of the Mandate of Heaven, and the contents of the pits would magically allow the Emperor to fulfill them in the afterworld as well (Wang 2012, 2015; Zhang 2016).

Taken together with the considerations we made about the "warriors", these ideas point strongly in the direction of a *unitary,* symbolical interpretation of the contents of all the pits—including thus the "warriors"—as "merely" single elements of a complex cosmovision of the afterworld of the first Emperor. This interpretation might be confirmed and better understood in the future if—as I strongly believe—further pits relating (don't ask me how) to the other two seasons may eventually be found.

We now turn to the problem of what the tomb proper might actually contain. First of all, we are sure (thanks to magnetic probes) that the tomb is located under the center of the mound and that it is a vast (around 80 × 50) underground palace. As a matter of fact, this would be consistent with a transformation of ancient Chinese burial practices and beliefs that would be brought to completion under the Han dynasty (Guolong 2005). This transformation was a long process, which occurred both architecturally—from Shang vertical pit tombs to underground burial chambers and apartments—and in the way the dead were viewed, from something like "frozen ancestors", venerated in specially dedicated temples and, as such, sources of political power for the living, to more present, active and possibly dangerous, ghosts, especially in case of violent death. At the same time, the burial equipment—which consisted chiefly of ritual vessels, bells and weapons—evolved, as well as becoming mainly symbolic. In this respect, the underground palace of the first Emperor is described by Sima Qian as a symbolic microcosm, a sort of miniature replica of the entire empire, where the great rivers of China are represented by channels of mercury and the vaults depict the heavenly bodies:

> With mercury they made the myriad rivers and the ocean, with a mechanism that made them flow about. Above were all the heavens and below all the earth.

Strange as it may seem at first sight, this description might be closer to the truth than one might be prepared to accept. First of all, anomalous concentrations of mercury have actually been found in the tomb's area, probably due to percolation from the mercury present in the underground palace (Chang 1985). Second, as far as the "celestial vault" is concerned, this kind of decoration is known to exist in several tombs of the Han period, thus only slightly later than Shihuang. An early example is the tomb from the beginning of the first century AD found at Jinguyuan, near Luoyang. The tomb has two chambers. The front chamber is decorated with swirling clouds (on the top, among the clouds, five round pebbles perhaps allude to the five elements). In the back chamber, twelve images correspond to months and refer to seasonal changes and to prescribed activities. The cycle is completed by the sun, the moon, the Five Elements and by two dragons ascending to heaven. In another tomb from the same period found in Xi'an, the "cosmization" is even more accentuated: the burial chamber is a sort of small capsule in which the vault completely engulfs the room with two concentric circles. The larger one contains the symbols of the Lunar Mansions and of the Heavenly Palaces, while the inner circle contains the sun, the moon, and a motif of birds flying amid clouds (Suhadolnik 2011).

All in all, if and when the Chinese authorities decide to open the Shihuang tomb it will be a momentous day for Archaeology worldwide. The interior of the underground palace might even turn out to be in good condition, even after 2200 years. Indeed, a (recently discovered) drainage system—essentially a underground "dam" of rammed earth—had been installed by the builders to keep the foundations of the mound and hence also the underground palace dry, and seems to have performed its function well to date. Sima Qian tells us that 700,000 men were sent from all over his empire to build the Mausoleum and, in spite of a tendency to exaggeration found in many ancient sources, in this case the figure seems likely (Rawson 2002). According to

Sima Qian, the tomb is protected by automatic crossbows traps, and a great number of sacrificed concubines had to accompany the dead inside, as well as those who knew the secrets of the interior palace (he goes on to say also that after the closure of the tomb, the mound was camouflaged by trees, but this is frankly difficult to believe, considering that only a few years later the mausoleums of the Han emperors would become powerful, unmistakable landmarks of power).

A final problem is that of the Emperor's body. Again, a series of interesting discoveries from the Han period may be of help in visualizing the situation. In particular, a fundamental discovery about funerary customs again emerges from the Mawangdui tombs. Besides a rich funerary equipment (which includes the already cited astronomical texts, and the funerary banners), the tomb revealed the exceptionally well-preserved body of Lady Dai. The body was treated with a solution containing mercury and is in an astonishing state of preservation, also due to the fact that the tomb chamber was perfectly isolated by layers of charcoal and white clay. Another impressive discovery, which furnishes us further clues, was made in 1968 in Mancheng County, Hebei. It is the tomb of Prince Liu Sheng (who lived some 100 years later than Shihuang) and his consort Dou Wan. This tomb is important for various reasons, not least for its impressive siting, which takes advantage of the shape of a double hill and is placed in a womb-like, natural environment (the orientation is almost precisely east-west). The tomb is accessed through a tunnel, at the end of which two side-chambers were located. One contained pottery jars, the other, the remains of four chariots and their horses. The main room contained a wooden tent-like frame covering two seats, in front of one of which vessels, lamps, incense burners and figurines were left. These "spirit seats" were clearly meant for the souls of the tomb's occupants, and so we find confirmation that the soul—or rather, the "lighter" part of it, originally worshiped in ancestor's temples—was considered as "dwelling" in the tomb together with the physical part, and essentially reunified with it. We also find a clear substantiation of the "ambiguous" state of the dead: the burial chamber was provided (don't ask) of a small toilet in the ground. Finally, and most surprisingly, the bodies were found wrapped in things which—in absence of better definitions—are called jade suits (Pirazzoli-t'Serstevens 2009; Rawson 1999). These jade suits are halfway between clothes and sarcophagi, halfway between symbolic suits of armor and funerary shrouds: Liu Sheng's, for instance, is made up of some 2500 small jade pieces, sewn together with gold wire. It is a long, laborious job to produce such a fine object, especially if it has to be done with the astonishing precision that can still be seen today.

Curiously enough, these jade suits were mentioned in texts of the epoch, but their existence was doubted by archaeologists until their actual discovery, and this is one of the reason why I personally believe that the interior of the Shihuang mausoleum may turn out to be not so very different from Sima Qian 's description.

3.3   The First Pyramid of China

There is no doubt that the first Emperor's building program at Lintong also served (besides his own dream of immortality) political and symbolical ends. First of all, the complex had to show explicitly that the deceased received the Mandate of Heaven rightfully. His conception of this mandate was fairly wide-ranging: as the earthly representative of the tutelary gods of heaven and earth, his figure was pivotal in the proper functioning of earth itself and of its inhabitants, and was to remain so also after death. The mound was essential to this aim: itself referred to as a "mountain", it was meant to be a replica of nature, over which the owner of the tomb exerted his power and control. At the same time, it was clearly artificial in that it was square-based and deliberately oriented to the cardinal points. The combination of these two elements— mountain-like shape and cardinality—helps to explain the origin of the shear idea of the mound, a problem which is usually overlooked in the literature (very few predecessors of Shihuang burial mound exist in China; among them, the royal tombs at Zongshan, where earthen mounds were used as earthworks to support temple-like structures over the tombs (Jie 2015; Loewe 1985)). The role of mountains—and that of the related concept of "cardinality plus a center"—was, indeed, fundamental in the creation of Chinese sacred geography. In this idealized view of the country, the heartland was bordered by sacred mountains. These mountains were: Mount Tai (Shandong) in the east, Mount Hua (Shaanxi) in the west, Mount Heng (Hunan) in the south, Mount Heng (Shanxi) in the north; furthermore, there was a "balance" or central mountain in Henan, Mount Song, along the southern bank of the Yellow River (Fig. 3.10).

Ideally, these borders would also be the borders of civilization, the boundaries of that precise place where the Mandate of Heaven was effective in eliminating chaos. This idea was already long-established and widespread by the time of the first Emperor, since the *Wuyue* ("Five Summits") doctrine is documented from the

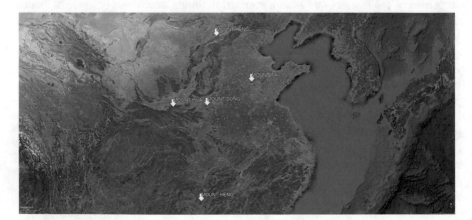

Fig. 3.10 Map showing the five sacred mounts of China. *Credits* Images courtesy Google Earth, editing by the author

Warring States period. This is something that Shihuang apparently knew very well. He devoted years to "inspection tours" to almost all the four corners of the empire, and made pilgrimages to Mount Tai and also to two others peaks in the east, Mount Xianglu (in Zhejiang) and—in his search for the land of the immortals, as we have seen—Mount Lao on Zhifu Island, in each case leaving hasty stone inscriptions documenting his passage (Kern 2005). On Mount Tai in particular, the connection of the sacred mountain with the divine rights of the ruler was made explicit with a solemn ceremony, the *Fengshan*. Complete details of the ceremony are unknown, but it was based on an altar mound covered with earth of five different colors, under which a jade tablet inscribed with ritual formulas was placed. I would not be surprised if some version of this ceremony replicated in terracotta and bronze were to be discovered in the Mausoleum's pits.

Before leaving this astonishing complex we must enlarge our view and consider the choice of its location. A first observation that can be made in this regard is that—in spite of what has been written in many sources—the position of the mausoleum is *in plain contradiction* with all the rules of Feng Shui. Actually, this is *not* surprising since, as we have seen, this doctrine had not yet been established in the Qin and Han period, and in this respect the case of Shihuang is crystal clear, as the most elementary Feng Shui rules would specify a mountain to the north and a river to the south, while here we have the exact opposite: Mount Li is in the south, the river Wei to the north (Fig. 3.11).

Therefore, the placement of the Mausoleum in the landscape—which of course was not random at all—must have followed *other* rules. What rules? First of all, the mound was positioned in a peculiar way with respect to the sacred peak of Mt. Li. The perspective is particularly striking if we observe the mausoleum from the

Fig. 3.11 Lintong, the siting of the Mausoleum of Shihuang with respect to Mount Li. View from the north-east. *Credits* Images courtesy Google Earth, editing by the author

north, as the line of sight to the Li mountains crosses the double peak called Wang Feng, an "armchair" ideally embracing the monument (Burman 2018). Secondly, the position of the tomb was ideal also with respect to the west, because almost due west of the mound we encounter the Qin capital, Xianyang. Xianyang itself was a seminal element of the building program of the first Emperor and of his propaganda; in particular, he ordered there the construction of replicas of the main palaces of the six vassal states. In this way, symbolically, the vital forces of the vassal kingdoms were transferred and imprisoned in a "microcosm" placed at the ideal center of the universe, the place where the Emperor—himself as center of the civilized world—resided. At least according to Sima, symbolism in the Qin capital went far beyond, to become an actual mirror of the Heavens. The first of the Emperor's palaces to be built there was named "Culmen Temple", to symbolize the celestial Pole. A direct road was built connecting it to the thermal springs at the foot of mount Li, the same place where—many centuries later—the Tang emperors were to build the Huaqing Palace, still visible today (on the road which takes visitors to the Tang palace, an impressive view of the Mausoleum complex serves to remind us who the first owner of these places was). Later, near Xianyang, a gigantic new palace was built, called the *Epang*, and this was also imbued with celestial symbolism. Today, only the huge rammed earth foundations of this building remain, but the project must have been truly impressive. The palace was connected with the capital by a bridge; cosmic symbolism was attached to this arrangement, with the river Wei playing the role of the Milky Way. Sima, in fact, tells us that the whole complex

Resembles the Elevated Passageway which extends from near the celestial Pole across the Milky Way to connect with Align-the-hall.

Here two Chinese constellations are mentioned: the Elevated Passageway is our Cassiopeia, while Align-the-hall is a part of our Pegasus. More precisely, Pegasus' core is formed by a huge and brilliant "square", and the asterism Align-the-hall was one of the "vertical sides" of the square, formed by the stars α and β Pegasi (Pankenier 2009, 2015). Indeed, with a touch of imagination, it is possible to imagine Cassiopeia as a cross-bridge over the arc of the Milky Way from the northern circumpolar region. The configuration was especially impressive at the end of October, when in the sky, viewed from the region of Xi'an in the third century BC, Cassiopeia with its characteristic "multiple arc" profile was seen culminating to the North while Align-the-Hall was culminating due south, with the vertical between its two main stars indicating the position of the cardinal point. Finally, at dawn, from the Epang terraces, the profile of Mount Li—emblematic of the emperor's tomb location—was visible on the far eastern horizon, creating a impressive sacred landscape of power.

In conclusion, then, the first Emperor's dream of immortality was made tangible and still stands today as one of the most complex projects a single human being has ever conceived. His idea of land marking the landscape of the heartland with a "mountain" sign would generate—over the course of the following two centuries—one of the most fascinating sacred landscapes human beings have ever created, that of the Western Han pyramids.

A sacred landscape we are, now, ready to encounter.

Chapter 4
Pyramids on the River

4.1 The Chinese *Age of the Pyramids*

The first Emperor died in 210 BC during a tour of Eastern China and was succeeded by one of his sons, Huhai. Huhai took the title of Qin Er Shi, that is, the second Emperor of Qin, following a nomenclature established by his father. According to Sima Qian indeed, Shihuang dreamed of an everlasting Qin empire, with a numbered succession of Qin emperors. Alas for him, it was not to be. The realm rapidly became unstable. China was fragmented again into separate kingdoms, and the second Emperor was forced to commit suicide by his minister Zhao Gao after only three years of reign. Visiting his tomb is a strange experience: it is just a very small pile of earth (difficult to call it mound) in a public park located not far from the Wild Goose Pagoda, in Xi'an.

This new "Warring States" period was not destined to last: in 202 BC, Liu Bang of Han (usually referred to as Gaozu) succeeded in unifying the state again, founding the Han Dynasty. Born to a peasant family during the late years of the Warring States period, under Qin rule, he was a low-ranking civil officer; apparently, one day he was in charge of accompanying a group of forced laborers to work to the Shihuang mausoleum, but they escaped. Fearing for his life, he started living as an outlaw. In 209 BC, the uprisings against the Qin broke out. Former kings of states conquered by the Qin began to rebel, and Liu Bang rapidly became one of the chiefs of the army of the Zhao kingdom; in 206 BC, his troops entered Xianyang. During the following, fragmented period, Liu Bang was dispatched to present-day Sichuan, from where he plotted his own conquest of the country, which ended with the establishment of the Han Dynasty in 202 BC. This was a long and prosperous dynasty, lasting from 206 BC to AD 220, although it was marked by a brief interruption from AD 9 to 23, when the usurper Wang Mang managed to ascend to power. This break had the important result that the Han rulers decided to move the capital to the east, to Luoyang, and this explains why the dynasty is formally divided into two periods, referred to as Western and Eastern Han respectively.

G. Magli, *Sacred Landscapes of Imperial China*,
https://doi.org/10.1007/978-3-030-49324-0_4

The founder of the new dynasty was of course faced with the problem of legitimating his own power. He thus devoted himself to the pacification and rationalization of the country. Initially suspicious about Confucianism, he gradually changed his mind and started recruiting Confucians for official state duties. Symbolically, legitimization of the dynasty clearly involved the adoption of the idea of a divine Mandate from Heaven. To express this concept through architecture, Gaozu embarked on a complex building program. This program included the foundation of a new capital, Chang'an, and the construction of a mausoleum that would explicitly display the supernal legitimization of the new order. The capital was built not far from Qin's Epang palace, to the south of the river Wei (Lewis 2005; Qingzhu 2007). The perimeter of the town was polygonal, not square, and historical sources report that the circuit wall was inspired by the stylized form of two constellations, the Northern Dipper, and a constellation called the Southern Dipper (essentially our Sagittarius), with the north celestial pole imagined to be inside the town (see Hotaling 1978 and the detailed discussion in Pankenier 2011). I am not fully convinced that this complex idea was at the origin of the capital's perimeter, but there is no doubt that the ideal identification of the emperor and of his residence with the circumpolar region of the sky became definitively rooted under Han rule. This, as we shall see, was to have significant consequences for the orientation of the Han mausoleums.

For his own tomb, Gaozu chose to follow the tradition started by the first emperor, and therefore his architects projected a tomb located under a huge burial mound, thus setting the benchmark for the Dynasty. However, unlike Qin's project, the location chosen was on the vast flatland located to the north of the river Wei, thus on the opposite side in relation to Mount Li. Most of the rulers who followed selected a building site in the same area, so that today the "pyramids "of the Western Han Dynasty create a fascinating sacred landscape. To understand the way in which this landscape developed, we shall now briefly visit all these monuments in sequence, since they are not very well-known to the general public and—especially on the web—rather ridiculous facts can be found about them, including an alleged "star project" for their placement referring to 10,000 years before their construction. Not even one of these tombs has ever been opened; however, their attribution (stated by a stela which bears the name of the owner) is rather sure (Zhewen 1993; Wu 2010). The stelae were indeed erected in the second half of the eighteenth century after a meticulous historical research; the identification of each tomb is therefore commonly accepted by archaeologists. The stelae also state the proper name of each monument, since starting with the Han dynasty each imperial tomb received a name comprising a word and the suffix -*ling*, tomb. The mounds have been preserved over the centuries and what is at risk today is essentially only their mutual inter-visibility, which however—as we shall see—is a crucial element of the sacred landscape they form. The area is well-covered by satellite imagery (Forte 2010; Kenderdine et al. 2012), with a resolution that is more than sufficient to measure the average sides and azimuths of the mounds; on the other hand, obtaining precise measurements in situ is problematic, since the sides are not well defined at many points (due to encroaching vegetation, cultivated fields, and washed-away borders). Therefore, although I have

Fig. 4.1 The Mausoleums of the Western Han emperors, numbered in chronological order. 1 Gauzu (Chanling), 2 Hui (Anling), 3 Jing (Yanling), 4 Wu (Maoling), 5 Zhao (Pingling), 6 Xuan (Duling), 7 Yuan (Weiling), 8 Cheng (Yangling), 9 Ai (Yiling), 10 Ping (Kanling). The mountain tomb of Wen is denoted by B. The position Q of the Mausoleum of Shihuang is also shown. *Credits* Images courtesy Google Earth, editing by the author

visited the vast majority of these monuments, the data which will be used here is based—as is the case for the rest of this book—on satellite imagery tools (Fig. 4.1).

The first tomb, that of Gaozu, is actually a twin project, made up of two huge, identical, rectangular mounds (dimensions about 135 × 168 ms) one for the Emperor and the other for Empress Lu. The original heights—as well as those of all the Han mounds—are difficult to determine, because the summits have deteriorated, but must not have been less than 50 m. The mounds are still today relatively solid, and this shows that they are actually "built" objects. In other words, they are not simple earthen mounds, but were erected by adding courses of "blocks" of strongly-packed rammed earth, a method which makes them homogeneous and very compact. All things considered, the word "pyramids" appears to be perfectly appropriate to describe them. As today, it was possible also in ancient times to ascend to the summit, and, indeed, scaling the tombs to offer sacrifices is documented in the Han annals (Fig. 4.2).

Thanks to my colleagues at XJTU, who helped me in locating the access to many of the Han pyramids, I have actually experienced the emotion of ascending Emperor Gaozu's mound and looking towards the area where the Han capital was once located, some 15 km to the south. The monuments are in a slightly elevated zone with respect to the town's area and must have both been clearly visible from the capital in ancient times, giving an impressive sense of power (Fig. 4.3).

The same holds for the tomb of the son of Gaozu, Emperor Hui (195–188 BC), who choose to build a rectangular mound that was virtually identical to the two of his parents, located about 3.5 km to the west of them. His complex includes three prominent satellite mounds, built for Empress Zhang, her father, the Marquis Zhang, and the Princess Lu.

Fig. 4.2   Ascending footpath to the summit of the Mausoleum of Emperor Gaozu, the first of the Han pyramids

Fig. 4.3   View from the summit of the Mausoleum of Emperor Gaozu. The huge, twin pyramid of Empress Lou is visible to the right

Hui's successor, Wen (180–157 BC) is an exceptional case. The Emperor is known for his reform of the empire, and for his frugality and concern for the people (Miller 2015). Wen's reign is anomalous in that he is the only Western Han ruler who did not build a burial mound. For his tomb, called Baling, texts say that Wen selected a mountain and had the funerary chambers hollowed out of the rock in it. To the north of the Wei, the river's floodplain extends for many kilometers, so that to remain near the capital, Wen was obliged to go to the south of the river. Apparently however, he excluded the chain of Mount Li. What remained to stand in for "mountains" was a river ridge to the southeast of modern Xi'an bordered by low hills which are barely distinguishable from flat terrain, although they do appear as hills from the opposite side. The Emperor selected one of these peaks, whose conical shape resembles a pyramid.

Wen's "mountain" tomb is fairly important in the history of Chinese tomb architecture since several elite and royal burials later imitated his idea, including, as we shall see, most of those of the Emperors of the Tang dynasty. Several theories have been put forward to explain the anomalous design and location. One such theory calls into question the fact that Emperor Wen could not have been legitimately buried in the region north of the Wei because he was the brother of the former Emperor and not his heir. Besides being dubious, this conjecture does not explain the choice of a natural mountain instead of a mound. Moreover, the fact that the tomb is under a natural hill has sometimes been explained as a sign of moderation and frugality, to avoid the huge expense and effort required for a mound-tomb. Again, the explanation appears doubtful, also because—as we shall see—the tombs of his wife and daughter are mounds, and almost as large as imperial ones. A sounder explanation may be that the unconventional choice was a way of conveying the identity and authority of the ruler, acting as a unifying pivot for the members of the royal clan. Symbolism, therefore, was the main reason for the unusual choices made by Wen. "Possession" of mountains was indeed a powerful symbol of kingship, exactly as it was for the first Emperor of Qin some 70 years before. Like Shihuang, also Wen conducted sacrifices in the mountains, and charged the local kings with conducting sacrifices in the mountains of their territories. Actually, we may also surmise that Wen (and also Emperor Xuan later on, see below) followed more closely the ideology of the First Emperor of Qin than their Han imperial predecessors did, locating his tomb south of the Wei, as the First Emperor did.

With the Mausoleum of Emperor Jing (157–141 BC), named Yanling, we return to the north of the river. The mausoleum set a new standard for the shape, which went from rectangular to square, and its dimensions (around 166 × 166 ms) would be chosen by almost all Emperors subsequently. Of course, these measures acquire sense if expressed in the Chinese unit of measure used at that time. The basic unit was the *chi*, which is relatively well known archaeologically due to the finding of several rulers of the epoch. Its average length is estimated at 22.75 cm, which gives an average length of 1.365 mt for its multiple, the *bu* (equal to six *chi*) (Bai 2015). Taking into account that an error in taking measures with satellite imagery has to be expected, the dimensions of the Yanling mound likely correspond to 120 × 120 bu, while the dimensions of the first three imperial mounds were most likely 100 × 120 bu.

Yanling is located to the east of the tombs of Gaozu and of Empress Lu, at about 5.50 km distance, and it is accompanied by a conspicuous satellite mound for Empress Wang. To the southeast of the emperor's tomb, a stone marked with a cross has been found; it was probably used as a station stone for making reliefs during construction, roughly in the same way as station stones are marked today (Jao 2007, 2013; Nanfeng 2013; Wuzhan 2011, 2017). In the same area, excavations revealed the traces of a complex of buildings probably devoted to the cult of the deceased. This corresponds to the transformation of the funerary customs occurred in the Qin-Han period we have already hinted at (Chongwen 2007). During the Zhou, after the initial mourning ceremonies, a memorial tablet carrying the name of the deceased was located in a specially devoted, "ancestral" temple for future sacrificial rites, which therefore were not held at the burial site. At the beginning of the Han rule, Emperor Hui ordered the construction of a temple in the area of the Gaozu mausoleum, which remained active as an ancestral temple of the dynasty; sources state that each 3 years all the memorial tables of the emperors were brought there and exposed following the *Zhaomu* (alternate, see below) order. The tradition of constructing ancestral temples to preserve the memorial tables will continue throughout the history of China, but with the building of Yanling each mausoleum complex started to include an imperial temple on its own, and the burial site started to become the main place for death rituals, regularly held on a daily and a seasonal basis.

Another important feature of Yanling is that the funerary pits of the mausoleum have been partly excavated, and arranged in a fascinating museum in which underground galleries bring the visitors into direct contact with their contents. The pits are ordered in parallel rows, orthogonal to the sides of the mound. They display the daily life of the Emperor's court, with various functionaries and servants, symbolically represented by miniature terracotta statues. For instance, the imperial kitchens are shown in detail, with stables for pigs and other edible animals. Many rows of warriors are also present, scaled around a third in size and relatively roughly-crafted (Fig. 4.4).

We are therefore certain that, as with the First Emperor tomb, funerary equipment of terracotta statues was also placed for the afterlife needs of the deceased in the Han imperial tombs (although—at least as far as we know—after Shihuang, life-size terracotta statues were never to be produced again). Recently, the discovery of terracotta statuettes in the pits of accompanying burials of some Han mausoleums, and of an entire miniature army in the tomb of a prince (Liu Hong, son of Jing's successor), has shown that this was not a privilege reserved for emperors.

The reign of Jing was followed by the very long sovereignty of Emperor Wu (141–87 BC). His tomb, Maoling, is exceptional, in being the largest Han imperial mound (around 248 × 248 ms, probably 180 × 180 bu), the highest one preserved (it is more than 45 m high) and also the westernmost. With a volume of more than 900,000 cubic meters, it is one of the most massive structures ever built by humans (by way of comparison, Khufu's pyramid at Giza is around 2.6 million cubic meters, and the Pyramid of the Sun in Teotihuacan, Mexico, is around 1 million). Maoling

Fig. 4.4   One of the burial pits at Yanling, with rows of miniature Terracotta "warriors"

is flanked by the highly conspicuous mound of Empress Li; at a close distance to the east, the tombs of various dignitaries are also visible. Among them, one is of special importance, that of general Ho Qubing, and will be discussed in detail later in this chapter (Fig. 4.5).

The successor of Wu, Zhao (87–74 BC) built his tomb to the east of Maoling at some 7 km distance. With the following emperor, Xuan (74–49 BC), we return

Fig. 4.5   The huge mound of the Maoling mausoleum

Fig. 4.6 The stepped mound of Empress Wang viewed from the top of Emperor Xuan's pyramid

south of the River Wei; today, Xuan's mound, together with its huge satellite, the terraced mound of Empress Wang, is set in a beautiful public park much frequented by kite-fliers. As far as the present author is aware, no explanation has ever been offered for the unconventional location chosen by Xuan. It may be noticed, however, that we have been handed down an image of him as being a hard-working Emperor, who grew up as a commoner, and was attentive to the needs of people. If this is true, perhaps an explicit reference to Wen's burial—who allegedly had a similar reputation—was intended in selecting the building site, and indeed a topographical connection appears to exist (see next section) (Fig. 4.6).

After Xuan, the emperors return to build to the north of the Wei river. The first three form a close group: Yuan (49–33 BC), Cheng (33–7 BC) and Ai (7–1 BC) are buried in very similar mounds (Cheng's stands out for the vast number of small satellite mounds which make, even today, a quite extraordinary sight). Finally, the last of the Western Han emperors, Ping (9 BC–6 AD), is buried in an over-sized (225 × 225 ms) mound located west of Yuan's. The origin of these huge dimensions are quite mysterious. Indeed, all the 3 monuments constructed earlier had standardized dimensions, and it is difficult to think that it was the emperor to ask for a greater mausoleum, since Ping had a short and tragic life, being poisoned to death by the usurper Wang Mang when he was only 13-year-old (Fig. 4.7).

Fig. 4.7   A mountain where there are no mountains: The impressive mole of Kanling, the pyramid of Emperor Ping

4.2   The Sacred Landscape of the Chinese Pyramids

The Chinese pyramids form a fascinating sacred landscape, which—as we shall see—did not evolve randomly. To study this landscape, we shall start with the orientation of each single monument. As we have seen, cosmization was a fundamental ingredient of any sacred landscape. In the case of a funerary monument, it assured that the deceased was "placed in the correct place" with respect to both the human and the afterworlds. Correct orientation of tombs has, of course, been fundamental in countless cultures; for instance, the sides of the pyramids of Giza (Egypt) are oriented to the cardinal points with astonishing accuracy, and the pyramids of China are no exception. Being rectangular or square objects, their orientation is easily defined as the azimuth (the angle with respect to the meridian; for a complete definition see the Appendix) of the two side bases closer to the north-south direction. These data are given in Table A.1, where all the measurable mounds (both the imperial and the satellites one) are reported; however, we only need to consider the imperial mounds, as the orientations of the satellites of an imperial monument are always very close and consistent with that of the main one. These orientations can be summarized as follows:

1. The mound of Gaozu, the first to be erected, is oriented at 167°. A close orientation is shared by its successor Hui, whose mound is also identical in shape and dimensions.

2. The third Emperor Wen marks the only departure from the burial tradition, since his "mound" is a natural hill; the orientations of the satellites of Wen will be discussed separately.
3. The fourth Emperor Jing returns to the burial mound tradition, but the orientation of his mound is cardinal.
4. With the two successors, Wu and Zhao, we see again a macroscopic deviation from true north, although the difference is less than that of Gaozu: the azimuths are 171 and 172 respectively.
5. We return again to cardinal orientation with Xuan (180°) and Yuan (179°).
6. A final oscillation occurs with Cheng (171°), with his successors Ai and Ping who both return to strict cardinality (180°).

We can divide these data into two, very neatly separated, families:

- Family (1): Precise orientation to the cardinal points, with errors not exceeding ± 1°.
- Family (2): Rough orientation to the cardinal points, with errors in relation to the geographic north of several degrees. These errors however are not randomly distributed: they are always to the west of north and exhibit a tendency to decrease in time from a maximum of 14° to a minimum of 8°.

These results clearly imply that two different methods of orientation were in use; indeed, the alternative would possibly be that cardinal orientation was sought with a single method which could cause errors of up to 14°. Apart from conflicting with the accuracy of the Chinese astronomers, this hypothesis is not feasible because we would have also data in the interval of errors between 1° and 8°. Moreover, a method implying that errors all lie on the same "side" is difficult to imagine.

There is no question as to the method used for the monuments of Family 1: it was clearly an accurate determination of the cardinal directions by means of an astronomical procedure. Astronomical methods for finding the meridian with good precision were readily available to the Chinese. First of all, circumpolar stars could be used, carefully observing their movements from a fixed observation point and establishing (with the help of assistants with poles and a level, artificial horizon) the maximal (east and west) elongations of a star close to the north celestial pole. Once these elongations are established, the meridian will be the direction between the observation point and the center of the corresponding arc, recorded on the ground. By the way, one can obtain in this way not only true north, but also the azimuths of the maximal elongation of the circumpolar star thus observed and measured, something which, as we shall see in a moment, was also of interest. Another star-based possibility is the use of the simultaneous culmination of two brilliant stars at due south, as suggested by Pankenier (2011). Establishing cardinality during daylight hours, on the other hand, could be done using the sun, marking with the help of a gnomon—fixed at the center of a circle—the directions of rising and setting on a flat horizon (Needham 1959). Once this has been done, drawing a line between the markers identifies east-west, and the perpendicular to this line through the center of the circle identifies the meridian.

Although solar methods are less accurate than stellar ones, both the stellar and the solar methods described—if used correctly—do not generally lead to errors greater than 1°. As already mentioned, besides explaining the orientations in Family 1, this also confirms that the orientations of Family (2) do not result from errors in determining true north, and another explanation must be sought. Since, as we have seen, some form of magnetic compass was very probably invented precisely under the Han Dynasty, one may suspect magnetic orientation. However, the magnetic compass during the Han—the compass spoon—was a sort of entertainment tool of court geomancers, requiring a delicate siting of the board, and was in any case a very crude instrument. It is frankly difficult to believe that it could be used to orient a building. In any case, I performed the test of correlation between the palaeomagnetic declination data and the data for the pyramids of Family 2 and the result is negative (for a detailed description of the method used in this book for testing palaeomagnetic correlations, see the Appendix). Strictly speaking, regarding Family (1), it should be noticed that, since the bandwidth of error in determining magnetic declination is relatively large, the pyramids of this family would fall into the realm of possible compass orientation as well. However, we must take into account the fact that the modern error in estimating the magnetic declination affects the positioning of the curve within the bandwidth, and it is not a physical error which can be associated with a measure point by point. Therefore, including both families in the same sample (as has been proposed in Charvátová et al. 2011) makes the magnetic interpretation inconsistent a priori. To explain this point in simple terms, consider, for instance, the royal mounds of Yuan, Cheng and Ai. They are very close in space (less than 7 km as the crow flies) and time. Yuan's mound is cardinally oriented; Cheng's mound—whose project was laid out on the terrain only 16 years later—is skewed 10° to the west, and finally, Ai's, designed on the ground 26 years later, is again cardinally oriented. Clearly, supposing that all the three have been aligned magnetically would correspond to a "crazy" behavior on the part of the magnetic declination in those years.

If magnetism is not a feasible solution, to find an explanation for Family 2 we have to have recourse again to Astronomy and to study the sky close to the North celestial pole in Han times. To do this, we must remember that, due to precession, the north celestial pole was in a dark region; it had been relatively close to the bright star Kochab, but in Han times started to move closer and closer to "our" pole star, Polaris. The target could therefore be the maximal western elongation—in actual terms, the distance in degrees—of one of these two stars from the pole. Polaris seems preferable because, if we follow its slow, apparent approach to the pole we can see that its maximal western elongation decreases in (rough) agreement with the gradual shift in the orientations of the Han mounds of Family (2), being ~13° in 200 BC, 12° 20′ in 100 BC, and 11° 40′ in 1 AD (Magli 2016).

To sum up, the likely explanation of both families of orientations is astronomical and linked with the most important region of the Chinese sky, the celestial Purple Enclosure. From the symbolic point of view, the identification of the Emperor with the celestial pole of the heavens immediately allows us to explain the orientations to true north of the monuments of Family (1), since the divine ruler was in principle meant to be located in the north, thus looking south over his kingdom. However, the whole polar region was identified as a celestial image of the Emperor's palace and court. In Han times, Polaris was not yet the bright star closest to the pole, Kochab being closer. Yet Chinese astronomers certainly knew (by direct observation) that Kochab had already reached the minimal distance (this occurred around the sixth century BC) and thus the pole was moving away from Kochab to Polaris. It is therefore quite conceivable that buildings were intentionally oriented to Polaris; orienting to the maximal elongation was the only way to do this, since, naturally, no orientation to rising or setting is possible for circumpolar stars.

Interestingly, the co-existence of two different customs of orientation—one exactly cardinal, the other skewed a few degrees either always to the east or always to the west—appeared in China long before the Han Dynasty. For example, during the Shang Dynasty, many palaces and official structures were built with cardinal orientation. At Xibeigang, Anyang, though, the royal tombs built during the late Shang period (around 1200 BC) and hundreds of their satellites all exhibit an eastward skew with respect to true north, which can be estimated at 10° (Loewe and Shaughnessy 1999). If the astronomical theory expounded above is adopted, then one could also explain this case of skewed orientation as pointing to the maximal elongation of a circumpolar star, perhaps Thuban (Didier 2009).

Finally, an explanation should be found for the different choices made for the "hand" of the elongation measured (either east or west with respect to true north). First of all, we notice that this choice is strictly unique for each cultural context (and indeed, uniqueness is a strong hint at symbolical content), but by no means unique in the sense of both space and time. Indeed, for instance, buildings in Erlitou (first half of the second millennium BC) exhibit a consistent skew to the west. Another example is the skew to the east exhibited by 938 Shang tombs at Yinxu, leading Keightley (1997, 1998, 2000) to suppose that they "perhaps venerated the north-east in some way". Probably, the solution is simpler and is, again, astronomical: seasonality. If tombs relating to the same cultural context were all planned on the ground in the same season of the year, then the choice of the "hand" for the elongation of the circumpolar star target would have come about naturally. For instance, consider the days close to the summer solstice in Han times: during the night, the western elongation of Polaris occurred in optimal visibility conditions, while the eastern elongation occurred during daylight and was therefore invisible. The reverse situation took place in the days close to the winter solstice.

All in all, we can say that the Chinese pyramids were "cosmicized" through a rigorous procedure of orientation. But this is only a part of the story, as together they formed a sacred landscape which did not evolve randomly. To study its evolution, it is essential to refer to a map in which the monuments are ordered chronologically. The distribution then reveals itself as very singular and, at a first sight, incomprehensible. In fact, one would expect the tombs to be placed in a linear succession from east to west (that is, at increasing distance from the capital) but it is not so, as "jumps" back and forth occurred. These jumps are, however, not random, but originate in a doctrine called *Zhaomu*. A contemporary source, the *Book of the Han,* dated around 111 AD (Wilkinson 2000), indeed mentions the use of this doctrine for the choice of tombs' locations of relatives. Essentially, the doctrine states that left/right (east/west) have to be alternately selected, so that, looking at a tomb, that of the successor will be to the left (west) and that of the second successor to the right (east). Not all scholars agree as to whether the method was effectively applied (Brashier 2011; Loewe 2016); however, from the satellite map it is clear that the alternate distribution was indeed applied to two triads of tombs: those of Gaozu, Hui and Jing (omitting the choice of a natural mountain made by Wen in between) and those of Yan, Cheng and Ai.

The application of the Zhaomu doctrine is a first proof that the necropolis of the Han emperors was conceived according to a general idea of how the landscape should develop. This does not mean, of course, that a master-plan was applied from the very beginning, but only that subsequent monuments were added in accordance with certain ideas and rules. The situation is actually very similar to what occurred in Egypt some 2300 years previously with the pyramid field of Giza, developed in accordance with a general topographical rule which was based with on inter-visibility with Heliopolis (as mentioned in Sect. 1.3). To study inter-visibility between the Han pyramids we will use the horizon formula, which gives the maximal distance at which an object of a certain height can be seen (details of this formula are given in the Appendix). To do this accurately, we would need the original heights of the monuments, which—as mentioned—is difficult to ascertain. For this reason, we shall take a cautious approach, considering the heights they reach today, which are of course either equal to, or less than, the original ones. It can then be easily seen that all the Han pyramids were inter-visible to each other from their summits. Indeed, the distance "as the crow flies" (that is, the length of a straight line) from the westernmost emperor's monument, Maoling, and the easternmost, Yangling, is about 35 km. Today, the highest point of Maoling is about 47 m above the ground and that of Yangling is about 25 m, giving a horizon visibility which is comparable to their distance. It follows that the Han monuments to the north of the Wei were *all* inter-visible at the time of their construction. In other words, they were all placed in such a way as to "speak" to each other along the visibility lines from their summits. Even today, notwithstanding haze and pollution, it is, in many cases, possible to appreciate from each mound the presence of at least the closest of the other monuments. The skyline was made even more fascinating by the presence of the satellite mounds. Indeed, the sides of the satellite mounds *never* align with the sides of the corresponding main mound, although they are oriented along the same azimuth. When I first noticed this, it seemed odd, because the orientation procedure must have been repeated for each one of them separately, while construction in alignment would have much simplified their planning. However, in this way, these mounds contribute to the creation of the skyline in a significant way, something they would hardly do in the case of side alignments with their principal counterparts.

Inter-visibility is not, however, the end of the story. In fact, there exist local topographical connections linking close monuments through visibility lines, which are clearly aimed at reaffirming the dynastic lineage. Two of these are particularly striking:

1. The project of Hui was explicitly, topographically connected to that of his parents.
    In fact, the north-west side of the mound of the Emperor is aligned with the south-
    east side of that of his mother Empress Lu. The result is quite spectacular: the
    twin mausoleums form a "sign of two mountains" on the eastern horizon of Hui's
    tomb (Figs. 4.8 and 4.9).

Fig. 4.8 A picture taken along the north side of the pyramid of Emperor Hui. In the foreground
(left), the huge bulk of one of his satellites, the pyramid of Marquis Zang. At the horizon, a "sign
of two pyramids", (outlined) those of the Emperor's parents Gaozu and Lou

Fig. 4.9 The alignment shown in Fig. 4.8. *Credits* Images courtesy Google Earth, editing by the
author

Fig. 4.10 The Zhaomu order and the alignment between the burial mounds of Yuan (centre) Cheng (right) and Ai (left). *Credits* Images courtesy Google Earth, editing by the author

2. The centers of the mounds of Yuan, Cheng and Ai are connected by an almost perfect straight line (Fig. 4.10).

Other geometrical relationship, of more difficult interpretation, also exist. For instance, there are two rather conspicuous, virtually identical satellite mounds to the north of the Jingling mausoleum, These two mounds are located on the same parallel, and at equal distances from the axis which passes from the summit of the main mound, so that the summits of the three form an isosceles triangle (Fig. 4.11).

Yet another example is the pyramid attributed to Empress Xu, wife of Emperor Cheng. It is the unique Empress' mound whose summit lies on the same parallel of the main burial. However, it is also the farthest (about 1.20 Kms) and the unique to be

Fig. 4.11 The Yanling Mausoleum complex: 1. Emperor's pyramid 2. Pyramid of Empress Wang 3. Station stone site 4. Twin satellite mounds. *Credits* Images courtesy Google Earth, editing by the author

endowed with many satellites on its own. Indeed, at least seven satellites are visible: a row of four to the immediate north, and other 3 further north, all of them oriented as the main, which in turn—as usual—is oriented as the Emperor's one. Perhaps the personal story of Emperor Cheng (who had many favorites, but apparently no heir born from any of them) is somehow reflected in this situation.

Finally, let us consider now the tombs located in the area to the southeast of Xi'an. Here, the situation is more complex, as the first monument, Baling, the tomb of Emperor Wen, is not an artificial mound. As mentioned however, Wen's funerary landscape comprises also two satellite burials: those of the Emperor's wife Empress Dowager Bo and of their daughter Empress Dou. Today, both monuments stand in a public park and are easily accessible. They are huge, almost identical, rectangular structures, almost the size of a imperial mound, located in the plain to the south-west of Baling. They are not, however, orientated as imperial mounds, because their orientation is not even approximately cardinal, being 23.5° east of north (Dou's longest side is rotated 90° with respect to Bo's). The solution is very probably that the tomb of Empress Bo was orientated towards the Baling peak, which is (barely) visible in direct alignment when looking from the summit, at a distance of some 3.6 km; the tomb of Dou was a replica of Empress Bo's, located, however, further south and therefore not explicitly oriented to the same target. Later, as we have seen, another Emperor, Xuan, chose the same area for his tomb. The diagonal of Xuan's mound neatly passes the apex of Baling, which was (again barely) visible on the horizon some 11.5 km away, probably to show an explicit connection between the two monuments. All in all, we see that also the burials in the area to the south of the capital were conceived in accordance with criteria governing sacred space. We can definitively conclude that the Western Han pyramids should be considered as an ensemble which stands as an imposing icon not only to each ruler separately, but to the dynasty as a whole.

With the Wang Mang break and the move to Luoyang, the tradition of building huge mounds starts to fade, and attention will move to the annexes to the tombs and, in particular, the so-called Sacred Ways, as we shall describe in the next paragraph. However, almost contemporary with the end of the construction of giant earthen mounds in China, a similar tradition starts in Japan, where enormous, keyhole-shaped mausoleums start to be built. These tombs are called *Kofun*. Among them, the hugest is the Daisen Kofun, which is commonly attributed to Emperor Nintoku (mid of the 5 century AD). Located in the area known as Mozu Tombs in Sakai, Osaka Prefecture, it is about 480 m long, 300 m wide at the base (the circular part is about 245 m in diameter) and more than 30 m high. The mound is surrounded by a large water moat and by many satellite burials. The cognitive aspects of orientation and landscape siting of these impressive monuments await for a comprehensive study; however, some examples of monuments astronomically or topographically oriented have been documented (Gowland 1897; Yoshitaka 2017). The problem appears to be quite challenging: for instance, as an experiment, I have found out that the transversal axis (the parallel to the baseline of the keyhole) of the Daisen Kofun, with an azimuth of 299° 30' and horizon almost flat, is very precisely oriented to the summer solstice sunset (the opposite orientation to winter solstice sunrise is less precise due to an horizon height of about 2°). This would accord well to the fact that in the few excavated keyhole Kofun the interior corridors, located under the round part, are usually placed along the transverse direction, as well as with the fundamental role of the Japanese Goddess of the Sun Amaterasu in the foundation of the imperial power. However, the baseline of another giant tomb, the Ojin Kofun, which lies not far, is not oriented in the same way and, to complicate matters, many satellites of these two main mounds share the same orientations, but many others do not.

4.3   From Huo Qibing to the Invention of the Spirit Road

Stone sculpture began to be practiced in China during the Han Dynasty, as first evidence of it comes from the reign of Emperor Wu, the possessor of the huge Maoling mausoleum (Paludan 1991). Wu (141–87 BC) was the seventh Han Emperor. His reign lasted 54 years, a length surpassed only by Kangxi of the Qing Dynasty. His figure is controversial: with Wu, China expanded its borders and rationalized the structure of the state, with a particular boost from the application of Confucian doctrines. Culture and arts were much encouraged, in particular music and poetry; many poems are attributed to the Emperor himself. Wu was, however, also a resolute military chief. For decades, the nomads known as Xiongnu had threatened the northern Han borders, and Wu dispatched his general Huo Qubing to solve the problem once and for all, expelling the enemies from the Ordos Desert and the Qilian Mountains. Another contribution by Wu to state organization is the introduction of the so called *era names*. Essentially, era names are labels used from that time by the Emperors to characterize certain years of their reign. The scope was essentially royal propaganda, since the names were chosen accordingly to main events or objectives (e.g. the era name Taiping—era of peace—was used by several Emperors). The count of the years was each time reset in accordance with the start of a new era (prior to the Ming dynasty, the Emperors would change the era name several times within their reign, while a unique era name would be chosen and associated with each single ruler under the Ming and the Qing Dynasties). Personally, Wu is one of the Chinese Emperors (together with Shihuang before, and with Jangli of the Ming Dynasty many centuries after) who became most fascinated by a yearning for immortality. At the same time, his despotism and harshness augmented; the Emperor became more and more obsessed by the afterworld and fearful of witchcraft-inspired attempts against him, leading to several collective executions, only to make public amends for his excesses in 104 BC with a document known as the Repenting Edict of Luntai.

As well as his huge tomb, which we have already described, the building program of Wu was quite extensive. Actually, the first Chinese stone statues we are aware of pertain to "microcosm" landscapes built for the emperors' delight. These kind of sacred landscapes, which will accompany us during the rest of this book, are symbolic representations of the world similar in many respects to those conceived some years before for the First Emperor's. Symbolism—as opposed, for instance, to portraiture—was therefore the inspiration behind Chinese stone sculpture right from the outset, probably associated with requirements of durability. In particular, Han sources relate that Wu embellished his hunting park outside the capital with an artificial lake inspired by lake Kunming, a beautiful, large lake in the Yunnan province. A bridge crossed the lake and at each end of the bridge a statue was placed, representing the Weaver Girl and the Cowherd, respectively (Schafer 1968; Howard 2006). These two characters appear in an ancient (at least sixth century BC) folk tale. The Weaver was a fairy in charge of weaving clouds; she fell in love with a man and married him, but her family sent her back to the sky. The Cowherd was allowed to ascend to the sky, but he could only gaze at his wife across the Heavenly River (the Milky Way); the two were in fact identified with two stars: the girl was identified with the star Vega, the boy with the star Altair (these two stars are actually seen in the sky as separated by the Milky Way). The two lovers can re-unite only once a year, on the 7th day of the 7th lunar month, by a bridge made out of magpies. The legend was widely popular, and a Han dynasty period song goes (Ping 2016; Wu 2010):

> Far away twinkles the Oxherd star;
>> Brightly shines the Lady of the Han River.
>> Slender, slender she plies her white fingers;
>> Click, click go the wheels of her spinning loom.
>> Her bitter tears fall like streaming rain.
>> The Han River runs shallow and clear.
>> Set between them, how short a space;
>> But the river water will not let them pass,
>> Gazing at each other but never able to speak.

Although it may seem incredible, the two sculptures survived the end of the Han dynasty and are still today located near the shores of the (long gone) artificial lake, in Doumen village. The statues are housed in two shrines, and the feast in honor of the re-joining of the two personages, the Qixi festival, still takes place as a sort of Chinese Valentine Day. The statues have quite surprising characteristics. First of all, they are both megalithic: the Cowherd is 258 cm high, the Weaver 228. Further, there is little attempt to depict real characters: the Weaver has a huge, long nose and resembles a rag doll (at least to a modern eye) but with a lonesome expression; the Cowherd is more dynamic but evokes non-human features. A second ambitious building project was the Jianzhang Palace, a new structure that Wu built as official residence to the immediate east of the capital. In the artificial lake of this palace, the architects placed a large stone fish resembling a dolphin or a whale, which has been recovered (it can be seen today in the fountain at the entrance to the Shaanxi Provincial Museum). The statue was probably part of a larger complex, also in this case somehow related to the spiritual world. Indeed, replicating landscapes—such as an existing lake—and locating them in places of power served religious and magical purposes, similarly to the palaces of the defeated states that Shihuang had rebuilt in his own capital.

Besides these scattered examples, we have a spectacular—although enigmatic—complete collection of statuary from Wu's times, recovered in the tomb complex of General Huo Qubing. This complex is a part of the landscape connected with the Emperor's own tomb, Maoling (it is today enclosed in the same archaeological site). Although aged only 24 on his death in 117 BC, Huo conducted the series of brilliant war campaigns mentioned above, and this is why his Emperor awarded him the honor of a tomb within the imperial complex. The mound appears today as a small hill, dwarfed by the huge mole of the Emperors' mausoleum. Sources report that it was intended to recall Mount Qilian, a high (5547 m) mountain in Gansu, the place of Huo Qubing's victories. Today, a pavilion has been constructed on the summit, and it is therefore impossible to ascertain if there was some special feature in its shape. In any case, the general's mausoleum really is a "one-of-a-kind": the whole mound was "covered"—it is difficult to find another word—by huge, sculpted boulders (Segalen 1912). Unfortunately, these sculptures are no longer in the original location, and their original position on the mound has not been recorded accurately; they have been collected and displayed in galleries along the flanks of the mausoleum. Therefore, we cannot know if they formed a meaningful iconography apparatus. In any case, the subjects represented can be divided into groups. A first group is that of animals: a tiger, an ox, horses, a toad, an elephant, two fish heads. Practically all of these animals exhibit curious features; in particular, they seem to look deliberately at the observer, with somewhat childish traits resembling human expressions. We are very far here from the solemn style of the terracotta warriors, but also from the close-to-nature style of the bronze waterbirds of the same mausoleum (Fig. 4.12).

Fig. 4.12 Tomb of General Huo Qibing, Maoling archaeological area: "Ox squatting at leisure"

Fig. 4.13 Tomb of General Huo Qibing, Maoling archaeological area: "Horse trampling a Barbarian"

Other statues represent dynamic scenes. The most famous is the so-called "horse trampling a barbarian", which is also the only one having a specific propagandist aim for the figure of the deceased, since the horse likely represents Huo Qubing's cavalry triumphing over the defeated Huns. The statue is also the only one which was found flanking the southern approach to the tomb, and is therefore also different from the others in function, being mainly ceremonial (Fig. 4.13).

The remaining "dynamic" carvings are quite mysterious; they apparently represent "monsters". In one, a human (or a giant) is hugging a small bear; a second shows a human-monster holding some sort of animal by the neck, while another animal near the back seems to emanate a breath of wind; the third human-monster has an oversize head and raises its right hand. These three sculptures clearly belong to the same "cycle" as they are essentially two-dimensional, being smoothly carved on the surface of huge slabs. It is difficult to appreciate the images without looking at them from above; this cannot be coincidental because, as Paludan (1991) remarks in her seminal book, the tomb of Ho Qubing bears all the signs of a most carefully thought-out project (Fig. 4.14).

I am, by the way, strongly convinced that we do not have the complete iconography originally present in Ho Qubing tomb. Indeed, especially on the north flank of the mound, many boulders and slabs can be seen which look like unfinished, or merely sketched out, sculptures.

Fig. 4.14 Tomb of General Huo Qibing, Maoling archaeological area: "Monster with small bear"

What meaning should we attribute to this complex iconography? One of the problems is that we do not have any parallels: no such thing existed before, nor anything similar later. As we shall see, funerary stone sculpture in China would rapidly become "official" in its aims and solemn in its subject matter. However, the basic idea was to remain the same: the sculpted creatures at Huo Qibing tomb represent a "afterworld microcosm", a land of spirits populated by beings who, once replicated in stone, would facilitate communication through magical channels. This idea would be elaborated in the centuries to follow, producing some quite stunning masterpieces. These began to be produced after the break that lasted from AD 9 to 23, and the consequent move of the Han capital in Luoyang. For reasons that are unclear, but probably having to do with an effort to strengthen political ties with the ruling family, Emperor Ming, the second Eastern Han ruler, introduced further changes in funerary customs (Wu 2009, 2010). He decided to introduce ceremonial visits of the court—including official occasions of meetings with nobles and local rulers—directly at the tomb of the deceased emperors. In this way, the tombs themselves became places for both the exercise of secular power (and the exhibition of such a power) and for ancestor worship. Accordingly, the need to extend tomb architecture for the use of living people arose (Steinhard 2002). This led to the construction of "sacrificial halls" to hold the ceremonies in, and to a monumental enhancement of the tomb entrance with a road lined with stone statues. In this way, a new, spectacular architectural element was invented: the *shen dao*, or *Spirit Road*.

It is easy to provide a first interpretation of the Spirit Road: since rituals involving the (living) Emperor involved a ceremonial approach during which the arrival of the monarch to the shrine was along a wide path lined with guards, so too a *magical* path lined with guards of various kinds (including animals and mythical creatures) was to be provided for his afterlife. Furthermore, since the Emperor was always imagined to be placed in the north, generally speaking the direction of the road towards the tomb was from south to north. Many other elements, however, played a role in the formation of the Spirit Road tradition which, with relatively few changes, will accompany us for the rest of this book, throughout 1800 years of Chinese architecture. The first element comes from Confucianism. One basis of Confucianism was filial piety, and constructing an imperishable, stone flanked tomb for a deceased was explicit proof of such piety: tomb monuments thus started to become status symbols, and the Spirit Road was widely adopted also for burials of the nobility, to the point that rules had to be drawn up about the quantity and subjects of the statues permitted, in order not to surpass the Emperor's ensemble in magnificence. In addition, the symbolic role of the road was to stress the connection between the two worlds, that of the living and the afterworld. In this sense, Shihuang's terracotta "warriors" can be considered as precursors of those of the Spirit Roads, at least if we accept—as I proposed in Sect. 4.3—that the terracotta "army" had a ceremonial, rather than defensive, function. Like the "warriors", the statues of the spirit roads also generally have a lost sight and are doomed to wait for eternity, but at the same time can fulfill their function of interconnecting the worlds 24 h a day in plain view, being incorruptibly made of stone. Stone in itself would thus become the main material for the graves, and stone stelae and statues a fundamental component of any tomb.

The general layout of the Spirit Roads rapidly became quite standardized. The road started with two paired *Que,* monumental towers, followed by animals and human statues, and ended with memorial stelae near the tomb. The towers were elaborate masterpieces, with meticulous carvings representing historical events or legends. Their origin lies in the wooden gate-towers used to mark the entrances to important buildings and also to towns; with the tombs they acted as symbols of a connection with heaven. In many cases, therefore, the four directional animals were engraved on them, "cosmicizing" the whole complex from the very start (Huo 2008; Wei 2017). The liminality of the place the visitor was about to enter was further stressed by the carving of a personage looking out from a half-open door.

The next section consisted of stone statues of men—representing guards of honor, military and civil officials—and animals, useful as guards in both the physical and spiritual world. As such, they could be depicted as real animals—like horses, elephants, rams and so on—or mythical beasts. A special role was played by tigers and lions, the latter being not native to China and known only through captive exemplars (the boundary between real and mythical animals was not very strict however, as felines with wings and horns also appear). Sometimes the names of the mythical creatures were explicitly mentioned, through a carving on the statue. These names are *tianlu, bixie,* or *qilin,* and correspond to imaginary animals—or pairs of male/female animals—used as talismans since ancient times and having the ability to bring good fortune and/or repel evils.

The last monuments placed on the road before reaching the burial mound were memorial stelae, used to show the titles and the merits of the deceased. Wooden stelae of this kind were already in use, so that those of the Spirit Roads represent the transposition into stone of an already existing tradition. The stelae had a sort of "crown" or head decorated with dragons, and a stone base. In particular, with the tomb of the royal official Fan Min (AD 205) in Sichuan, we encounter the first example of a *tortoise stela*. Essentially, a tortoise stela consists of a rectangular monolith mounted in an square cavity, appropriately carved, on the top of a monolithic, sculpted block. The block represents a tortoise, and therefore the effect is that of a giant turtle carrying a huge stone on its carapace (the tortoise's head sometimes has lifelike animal features, or may have the features of one of the mythical beasts) (Fig. 4.15).

Tortoise stelae will become during the centuries also the landmarks used by Emperors to state their interest and devotion to special places. One example of these landmarks is visible in the Confucius Temple of Qufu (we shall visit it in the next Chapter) another is in the Beilin, the Museum of the Forest of Stelae in Xi'an. Anyone who has visited it will have had the breathtaking experience of seeing scores of such tortoise stelae carved over 1800 years of Chinese history. In this book we shall also visit the little-known, most massive tortoise stela ever conceived, which remained at quarry block stage, due to the sheer immensity of the project.

Fig. 4.15 Tortoise stelae at Confucius' temple, Qufu

To conclude, I can say that walking along a spirit road is, even today, a strange and awe-inspiring experience. The statues are always twinned and face each other across the way, but appear completely uninterested in looking at the visitor, who feels quite small when faced by the their solemnity and, in many cases, overwhelming hugeness.

After sunset, in the silence, yes, just for a brief moment, it really looks like they are preparing to mount the night guard against the evil spirits of the night.

Chapter 5
The Golden Age

5.1 From Pyramids to Mountains

The decline of the Han empire occurred in the second half of the second century AD. China entered a turbulent period, in which the country was divided into three states. A first unification occurred under the Jin dynasty, a second, partial one during the period of Northern and Southern Dynasties (420–589 AD), in which China was divided into two kingdoms. Of these, the Southern one (with capital Nanjing) continued the tradition of tomb mausoleums with statuary, mostly located in the eastern suburbs of the capital. The short Spirit Ways of these tombs always included two huge winged and agile felines, which still stand out as their most distinctive elements today. With the north finally conquering the south, the country was unified again under the short-lived Sui dynasty (609–618 AD) and finally commenced a new period of splendor with the Tang Dynasty (618–907 AD), founded by Emperor Gaozu.

Gaozu, born Li Yuan, was Governor of modern-day Shanxi. Championing the revolt against the Sui, he was able to settle China under his control with the help of his second son Li Shimin, the future Emperor Taizong, to whom Gaozu passed the throne while still in life, in 626. Taizong was the true stabilizer of the dynasty and the initiator of a period of more than a century of stability. Under him, the Tang armies defeated the Western Turks and annexed their Khaganate (kingdom), which included vast regions of central Asia, stretching as far as the Caspian Sea. Taizong is indubitably one of the most outstanding of Chinese rulers (Fitzgerald 1933). Under Taizong, the Tang Dynasty reached its apex: considerable development took place in art, literature, and science. The capital of the empire, Chang'an, became the icon of the new era. The Sui rulers had already refounded the city slightly southeast of the original Han site. Under the Tang, Chang'an became one of the largest—probably *the* largest—cities in the world (Schinz 1996; Turnbull 2009). To have some idea of its magnitude, one can visit the imposing *Ming* walls of Xi'an, built in 1370. These walls stand virtually intact, encircling the rectangular area of the Ming era city for some 12 km (one can ride round the whole perimeter on a mountain bike). They

G. Magli, *Sacred Landscapes of Imperial China*,
https://doi.org/10.1007/978-3-030-49324-0_5

are more than 12 m in height, and reach 15 m thickness at the base. Climbing up the magnificent south tower and looking south, one can try to imagine the size of the much larger *Tang* city, whose walls extended for some 8 × 10 km and whose estimated population was half a million people. Tang Chang'an decayed rapidly at the end of the Dynasty, but an almost intact monument on the horizon reminds us of the greatness of the Dynasty: it is the Giant Wild Goose Pagoda, built in 652 AD, located in the southeast sector of Chang'an, well out of the Ming walls. The monument was built to house sacred materials that had been brought from India by a distinguished scholar of the history of Buddhism in China, a monk called Xuanzang. He made a 17-year-long journey to India, returning to the capital with a variety of images, books, and Buddha relics. Under the auspices of the state, he founded a Buddhist temple—the Daci'en—in memory of Empress Zhangsun, the wife of Taizong and mother of his successor, Gaozong. He also obtained permission to build a pagoda to stand as witness to the sanctity of the place. The original project probably already rose to an height of 60 m with five storeys, while the version we see today—restored under the Ming dynasty—is more than 64 m high and has a total of seven storeys (it is fully accessible with easy staircases) (Fig. 5.1).

Under Taizong, a new architectural paradigm was introduced for the emperors' tombs (Zhou 2009). Indeed, sources report that the he ordered the construction of his father's mausoleum according to the Han burial mound traditions, but for his own tomb, he was inspired by Baling, the tomb of Emperor Wen of Han and the only Han ruler who built his tomb under a natural peak. Gaozu's tomb was thus built on flatland and is characterized by a huge mound, but the Taizong tomb, Zhaoling, was located under a mountain, Mount Jiuzong, to the north of the Wei flood plain. The peak has a very distinctive profile, and ancient sources say that it was sufficiently high as to be seen from the capital, as a little heap at the horizon. Of course today it would be impossible to have this view, but it may well be that it was possible in ancient times. Indeed, the relative height of the peak with respect to the town is around 600 m and the distance from the capital is about 55 km, which is within the theoretical visibility given by the horizon formula (around 88 km). This theoretical value, of course, does not take into account local conditions, as visibility in the area has always been affected by the natural loess dust burden of the atmosphere in almost all seasons. However, on very clear days, the view must have been possible and, interestingly enough, satellite imagery shows that a line (unobstructed in antiquity) from Xi'an center (Bell Tower) to the highest point of the mountain cuts the huge mound of the Yiling Mausoleum of Emperor Ai of the Han dynasty very neatly—this was perhaps used as a guide to the eye.

Taizong's mausoleum was conceived as a walled enclosure—in a sense, as a replica of a town—with four accesses located at the cardinal points and guarded by stone lions. A Spirit Road leads to the mountain and included six beautiful stone reliefs representing the steeds of the Emperor (today 4 are exhibited in the Stelae Forest Museum of Xi'an and 2 at the Museum of the University of Pennsylvania, USA) (Ferguson 1931). In selecting a mountain for his tomb, Taizong initiated a tradition which was to be followed by almost all the emperors of his Dynasty, who chose different peaks of the same mountain range. As a matter of fact, the successor

Fig. 5.1 The Wild Goose Pagoda, Xian

of Taizong, Emperor Gaozong, is—together with his wife Wu Zetian—the possessor
of the most magnificent of these tombs (Fig. 5.2).

The reign of Gaozong is unique in the history of China. Gaozong formally ruled
from 649 to 683, but in 665 he fell severely ill and imperial powers were exercised
by his second wife Wu Zetian, a former concubine of Emperor Taizong. In late
683, Emperor Gaozong died. Their third son Li Zhe was enthroned as Emperor
Zhongzong, but his mother remained regent. As soon as Zhongzong showed signs of
disobedience, she exiled him and had her youngest son Li Dan enthroned as Emperor
Ruizong. Wu Zetian thus remained the true ruler, and finally—in 690—she decided
to establish, the only woman in the history of imperial China, her own dynasty, that
she called Zhou. This formal break in the Tang rule lasted for 15 years: in 705,
Zhongzong managed to return to the throne, and Wu Zetian died shortly thereafter.

Fig. 5.2 The Tang mausoleums, numbered in chronological order. 1. Gaozu (Xianling), 2. Taizong (Zhaoling), 3. Gaozong (Qianling), 4. Zhongzong (Dingling), 5. Ruizong (Qiaoling), 6. Xuan-zong (Tailing), 7. Suzong (Jianling), 8. Daizong (Yuanling), 9. Dezong (Chongling), 10. Shunzong (Fengling), 11. Xianzong (Jingling), 12. Muzong (Guangling), 13. Jingzong (Zhuangling), 14. Wenzong (Zhangling), 15. Wuzong (Ruiling), 16. Xuanzong (Zhenling), 17. Yizong (Jianling), 18. Xizong (Jingling). *Credits* Images courtesy Google Earth, editing by the author

One of the most important features of the reign of Wu Zetian was the favoring of Buddhism over Taoism and Confucianism, already initiated by her husband with the construction of the Daci'en temple, which caused much dismay among Confucian scholars. One of the most astonishing Buddhist masterpieces in China, the Fengxian Caves of the Longmen complex stand as a testament to this age (Fig. 5.3).

The Longmen Caves are a series of caverns (but they might be better defined as huge niches), excavated 12 km south of Luoyang, along the course of the river Yi. The grottoes are excavated into the limestone cliffs of the riverbed. The sheer number of statues is impressive (some estimates amount to several tens of thousands) distributed in several different shrines; carving started during the reign of Emperor Xiaowen of the Northern Wei dynasty (who made Luoyang his capital in 493) but continued well after the Tang and up to the Northern Song period. The most astonishing cavern is the one completed in 675 under Wu Zetian, the Fengxian. Along the wall, the rock has been carved to create nine huge statues, the largest being the central Vairocana (celestial) Buddha which is more than 17 m high. Buddha is flanked by two disciples and two monks on each side, and protected by a Lokapala (celestial guardian spirit) holding a pagoda in one hand, and a Vajrapani (protector spirit) to his side. Today, the sculptures are in the open air, and it is difficult to believe—as sometimes claimed—they were originally carved inside a (today ruined) grotto. Thus, their Buddhist message—but of course, also their political one, given the endorsement of Buddhism by the ruling family—could be seen as a spectacular welcome by all people traveling to Louyang from the south along the river (Fig. 5.4).

Fig. 5.3 The Fengxian cave of the Longmen complex, Luoyang

Fig. 5.4 The Longmen complex, Luoyang, as seen from the river Yi

Fig. 5.5 Qianling. The sacred way looking towards the mountain

Fig. 5.6 Qianling. The *que* hills seen from the summit of the sacred way

Gaozong and Wu Zetian are buried in the most magnificent of the Tang Mausoleums, Qianling. It is located at the foot of Liangshan Mountain, not far from the Taizong mausoleum, which is plainly inter-visible. The main peak of the Qianling mountain is cone-shaped; it actually resembles an enormous artificial mound to the point that it is sometimes presented as such in naive publications (Fig. 5.5).

It is interesting to try to repeat the mental exercise that the imperial architect must have done when he first visited the place and imagined the structure of the royal tomb. Qianling is, in fact, a masterpiece in its combination of natural and man-made features. This is already apparent from a distance, when the mausoleum reveals itself in relation to the *Naitoushan* (Nipple Hills). These two hills stand to the south of the main mountain, and were used as natural *que* pillars, their shape being enhanced with man-made towers (Fig. 5.6).

After this spectacular entrance, the visitor meets the Spirit Road, lined with numerous stone sculptures. The path is opened by a pair of ornamental columns with octagonal section. Next, the visitor encounters auspicious winged horses. Five pairs of stone horses with grooms follow (only partially conserved); next, ten pairs of guards with helmets and swords. At this point, the road reaches a platform holding two huge stelae (Fig. 5.7).

The one to the east—called the Uncharactered Tablet—is more than 6 meters high and was originally left blank under Wu Zetian's will, as she wanted later generations to evaluate her life (actually, many later inscriptions have been carved on it). The

Fig. 5.7 Qianling. Uncharactered stela; rows of stone ambassadors can be seen in the far left

west stela, although huge, is not monolithic. The inscriptions on it celebrate Emperor Gaozong achievements. Next, we encounter rows of life-sized stone figures representing nobles or foreign dignitaries (today most are headless). The final part of the path is introduced by a pair of stone lions and leads to the huge natural mound under which the underground palace was placed. As with all others imperial Tang tombs, the grave has never been opened; we can, however, get a glimpse of how it might look by visiting the satellite tombs which have been excavated in the necropolis associated with the main mausoleum. In particular, an important tomb is the one belonging to the crown prince Li Xian, located some 2.5 km to the south-east of the Qianling peak (Fong 1984; Eckfeld 2005; Wu 2010).

The life of Li Xian (655–684) is dramatically interconnected with the desire for power on the part of his mother, Wu Zetian. Indeed Li Xian—second son of Gaozong and Wu Zetian—was a legitimate heir during the years in which she managed to retain control of the empire by various means, and therefore he was potentially a threat to Wu Zetian herself. Eventually, we do not know whether rightfully or not, Li was accused of betrayal, and degraded to a commoner. First he was exiled and subsequently forced to commit suicide. His memory was rehabilitated after his mother's death in 705, and his body was reburied in the royal tomb constructed for him in 706. The tomb is set along a straight axis introduced by a pair of watchtowers and a pair of stone rams. A wall of rammed earth encloses the inner compound at the center of which a square mound denotes the presence of the underground apartments. These are organized along a straight axis for a total of 71 m length. A visit to the interior is an amazing experience, as the entire tomb is a gallery of painted art: the walls were painstakingly covered by a clay mixture, particularly suited for painting, and magnificent murals adorn the corridors (the colors were obtained by a variety of pigments and minerals). At the entrance, a large descending ramp leads underground with the ceiling increasing in height. The walls show lively scenes: hunting (east wall) and polo playing (west wall). A middle section follows, with images of officers and honor guards (in the ceiling, a series of light wells can be seen, whose function is—at least to me—very unclear; one of these was likely used by robbers). On the sides, there are niches which were used to store ceramic statuettes (Keightley 1991). Entering the next section gives the illusion of entering in a garrison, with the depiction of racks containing halberds. The corridor, guarded by painted personages, becomes horizontal and leads to a first chamber decorated with scenes and members of a noble court (eunuchs, musicians and dancers). The final sector, originally closed by a stone door, leads to the coffin chamber. The huge stone coffin (made out of joined slabs, otherwise it would have been impossible to carry it downstairs) is decorated with architectural details and human figures as well. Besides the coffin, the chamber contained two stone epitaph tablets of the deceased. In spite of the looting, hundreds of pottery statues were found (mostly broken) in the niches; they represent a variety of subjects, testified to also by the excavation of other tombs of the Tang period: most figurines represent singers, dancers and dignitaries; there are also equestrian figurines, domestic edible animals, horses and camels. Protective figurines included also mythical flaming beasts and a couple of heavenly kings.

All in all, then, Li Xian tomb's iconography replicates contemporary life and entertainment at court on the walls and in ceramics, but also includes clear references to the afterworld. To assure that the deceased is correctly placed in the cosmos, the ceiling of the tomb appears to have been decorated with a depiction of the sky (the stars were made out of gold leaf, which was looted by robbers) as in the Han (and probably Qin) tradition. One can only have a glimpse at what the opening of the Qianling underground palace may eventually reveal in future in terms of artworks.

During the Gaozong and the Wu Zetian rule, two other important mausoleums in imperial style were built. The first is Gongling, the tomb of the crown prince Li Hong (who predeceased his father in 675) located in Yanshi, Henan. The tomb is provided with a huge square mound around 140 ms side base. The other is Shunling, the tomb of Wu Zetian's mother, Lady Yang (located close to Xi'an International Airport). Lady Yang was originally buried in a standard noble tomb, but in 684 the empress raised her mother to royal status posthumously and in 689, on the way of proclaiming emperor herself, to imperial status. As a consequence, the tomb was enlarged and the Spirit Road was elongated and increased two times. The original tomb is a small mound with an access way guarded by statues of officials, rams and tigers. The royal Spirit Road is attached to the previous one and features seven further pairs of officials. When the project was enlarged again to become a imperial mausoleum, a walled precinct guarded by pairs of stone lions and a pair of earthen Que at the southern entrance were added. The enlargement of the Spirit Road was never finished, but beautiful, huge statues were put in place. In particular, a pair of monolithic statues of winged, propitious animals can be seen, each almost 4.15 m high and more than 4 m long, among the largest statues ever sculpted in China (Fig. 5.8).

After Qianling, as many as sixteen other Tang emperors' mausoleums were constructed. Some of these are of difficult access, to the point that in some cases

Fig. 5.8 Shunling, the Sacred Way

the statues emerge from cultivated fields; farmers have been used to seeding the terrain avoiding the sculptures for centuries. The Spirit Paths are standardized: from south to north columns, winged animals, ostriches, guards of honor, horses with grooms, civil and military officials, gates guarded by lions. After proclaiming its connection with the afterworld through the winged animals, therefore, the standard Tang Spirit Road focused on real court life. However, the expression of the human figures remained solemn and distant to the observer, and the animals were depicted in static poses. The necessity of choosing peaks suitable for the construction of the spirit road led to the spread of the mausoleums along the Guanzhong Plain, extending some 150 km from the easternmost, Qianling, to the westernmost, Tailing (Gong and Koch 2002). In my view, they form a quite peculiar case of sacred landscape, since their "built" features are not inter-visible at all. Looking from the plain on a clear day, however, the long range of hills can be seen and so—provided that one *knows* that, but this knowledge was widespread—many mausoleums are actually seen at the same time. The specific orientation of each one can be defined by measuring the Spirit Path, which is always straight and—at least in most cases—does not depend on local topography. It is actually possible to measure the axes of thirteen imperial tombs, as well as those of the two imperial-style mausoleums. The azimuths, looking along the path in the direction of the tomb, are given in Table A.2. The following observations can be made:

1. The first mausoleum, as already mentioned, is actually a mound in Han style, and it is oriented cardinally.
2. The orientation of the Spirit Path of the first mountain-based mausoleum, that of Taizong, is unique in that it is directed southward (10° west of south). In other words, this is the only one of these monuments that is approached from the north. There is probably a topographical reason for this, as the mountain rises very steeply on the plain to the south, while on the opposite side the approach is from a high plateau, with a sort of smooth valley, in which the Spirit Path was built. However, a concomitant (not mutually exclusive) reason may exist. Indeed Taizong closely stressed his dominion over the Turks' culture of the north, to the point of assuming the title of "Heavenly Kaghan" in order to legitimize himself as a steppe Khan. Whatever the reason, this orientation was no longer used; in other words, all other Tang Spirit Paths run from the plain towards their peaks to the north.
3. Two mausoleums—Jianling and Tailing—have orientation at 348°. Examination of satellite imagery shows that topographical considerations imposed this choice. The Jianling mausoleum is unique in that it actually has two, roughly parallel Spirit Paths (359° and 348°) which develop along two parallel natural reliefs that approach the mountain. Apparently, the requirement of symmetry for the approaching path obliged the architect to build two of them. The case of Tailing is similar, as the builders were compelled to conform the symmetry in relation to the surrounding hills.
4. The remaining mausoleums all exhibit orientations between 350° and 357°.

The fact that these last orientations were intentional is beyond doubt, as can be seen, for instance, by inspecting the left (west) side of the Spirit Path of Qianling, which is still supported by a huge retaining wall built precisely in such a way that the path over it could have the chosen orientation. To understand the skew to the west of north, observe that the azimuths are similar—but generally speaking, greater and therefore closer to true north—to those of the Han family's two mounds which, as we have seen, are probably oriented to the maximal elongation of Polaris. Therefore, we are left with the same option for the Tang mausoleums. Actually, during the Tang dynasty, Polaris' elongation was ~10° in 600 AD reaching ~8.5° around 850 AD, in essential agreement with the data, which show the same tendency to decrease. Interestingly, in the same period a very precise knowledge of the circumpolar region and of the position of the north celestial pole is also confirmed by the first known complete star chart of China, the Dunhuang map. This map, reasonably dated to the Early Tang, includes more than 1300 individual stars divided into 12 sections, plus the circumpolar region; in particular, the last section is the most detailed, displaying 144 stars. In the chart, Polaris is singled out and named individually (Bonnet-Bidaud et al. 2009; Schafer 1977).

Similarly to the case of the Han mausoleums, the alternate possibility of a magnetic orientation for the Tang tombs can also be excluded. Indeed, following the method described in the Appendix, the correlation coefficient of azimuth vs. magnetic declination can be calculated to be very low (R = 0.29). Furthermore, all deviations measured are west of north, while the magnetic declination changed sign in that geographical area around 700 AD (see again Table A.2). As a consequence, in case of correlation, the orientations should show a tendency to switch from values west of north to values east of north, something which is not observed in the data.

5.2 Buried in the Homeland

At the end of the ninth century, the Tang Dynasty was overwhelmed by a series of revolts by regional warlords, leading to a historical period so fragmented that it is usually referred to as Five Dynasties and Ten Kingdoms (907–960). This period formally ended with the establishment of the Song dynasty on the part of Zhao Kuangyin.

Zhao was born in Luoyang, and was the son of an army officer. He grew up to be a formidable horseman and fought in several of the wars involving various states and ephemeral dynasties during the turbulent period of his youth. His ascent to the throne in some respects recalls coups that occurred in the history of Imperial Rome. Apparently, a diviner claimed that he saw two suns fighting, meaning that the Mandate of Heaven was in a turbulent state and had to be transferred (of course, to Zhao Kuangyin). The army then acclaimed Kuangyn as Emperor, and he, of course reluctantly, accepted, acceding to the throne as Emperor Taizu. Immediately afterwards, he devoted himself to reuniting most of the country under his rule. However, the Song Empire remained bordered by independent rulers who asserted their own dynastic lineage. Indeed Manchuria, Mongolia and parts of Northern China remained independent and ruled by the Liao Dynasty (907–1125) while Gansu, Shaanxi, and Ningxia were governed by the Western Xia dynasty (1032–1227). Furthermore, the Song period is divided by historians into two parts, called Northern and Southern Song. The reason for this is that during the Northern Song (960–1127), the capital was in Kaifeng, and the Dynasty controlled most of what is now Eastern China. The Southern Song period (1127–1279) followed the loss of control of the northern part in favor of yet another dynasty, the Jin. During this second period, the Song court retreated south of the Yangtze and moved the capital to Hangzhou.

In spite of its relative military weakness, the Song rule actually produced a splendid flowering of Chinese civilization. This had already started with Emperor Taizu, who launched a decisive return to tradition, at the same time favoring a meritocratic approach through the enhancement of the function of the imperial examination system for state recruiting and encouraging academies and culture in general. In particular, the Song re-visitation of traditions led to the reformulation of Confucianism as Neo-Confucianism and to the birth of archaeology, to the point that many ancient stone inscriptions have been passed down to us only through Song copies. At the same time, many scientific breakthroughs took place, as well as an expansion of maritime trade and industry. Moreover, for some reason (perhaps due to the will of the emperors to display more power and wealth than they actually had), the Song Dynasty marked one of the most outstanding moments for megalithic sculpture and construction.

The prime example of megalithic Song structures are the bridges (Bigoni et al. 2017). To understand these astonishing engineering feats, a technical observation must be made. The use of single blocks of stone as beams, or lintels, becomes increasingly dangerous when the span of the stone between the supports increases. This is due to the weight of the stone itself, to the weight which may be applied to it, and to the possible existence of internal fractures and cracks. For this reason, although in ancient times single blocks of stone of lengths greater even more than 20 m were quarried and used in plain masonry or as standing stones (like, for instance, many Egyptian obelisks), the use of single blocks as beams (thus supported only at the two ends) was usually relatively limited, and methods to relieve the weight from them were exploited. Actually, it seems that unsupported spans seldom reached 7 m, and this limit was more or less heuristically established by many ancient architects independently. This, however, was not the case for the Chinese architects of the bridges built during the Song Dynasty in the Fujian region. Here, Needham recorded the astonishing existence of beams which he estimated to be up to 21 m in length and more than 200 tons in weight, attributing them to a single "great master of primitive bridge building". We shall probably never know the name of this master, but his bridges still exist today (so that they are not so primitive, after all). In spite of the fact that they have been heavily over-built, it is easy to see their original construction technique: they are just made of huge, longitudinal stone beams spanning from pier to pier. The beams are made of granite, while the piers are built of solid, dry masonry. One example is the Luoyang Bridge, built around 1150 and still in operation for light traffic. Here, a standard beam is about $9 \times 0.7 \times 0.5$ m for a weight of more than

Fig. 5.9 Luoyang Bridge, Hangzhou. Notice the monolithic stone lintels under the modern balustrade

120 tons; another example is the Jiangdong Bridge on the Jiulong River. Here, five ancient piers are still in place, with granite beams as long as 14.60 m, while two remaining granite beam pieces confirm that blocks spanning around 18.60 m were also used (Fig. 5.9).

Another example of the Song architects' familiarity with huge stones can be seen at Qufu, in Shandong. Here a complex sacred space was formed over the course of the centuries, for very important reasons. First of all, Qufu is the birthplace and the burial place of Confucius. To be precise, Confucius' parents were from Qufu, and Sima Qian relates that one day they went to pray at Mount Ni, a lowish (300 m) but distinctive mountain which is located some 25 km south east of Qufu, while his mother was pregnant, and that the philosopher was born on the mountain (today, the Nishan hill is the site of a temple and of a Confucian Academy). In addition to being sacred to the Confucians, Qufu is traditionally considered the birthplace of the Yellow Emperor and the burial place of his son, the second mythical emperor. The sacred space at Qufu thus comprises two main elements: *San Kong*, "the Three Confucian Places", and the second Emperor's Mausoleum called *Shou Qiu* (Fig. 5.10).

The first element of the Confucian complex is the huge temple of Confucius, actually the largest and most renowned one, which occupies a substantial part of the walled, ancient city of Qufu along the east side of its meridian axis. The temple we can see today is the result of more than 2300 years of evolution, since the house of Confucius began to be the subject of veneration rapidly after his death. Later on, many emperors paid homage and made sacrifices here. A major restyling occurred

Fig. 5.10 The sacred landscape at Qufu: 1–2–3. Confucius Temple, Mansion and Cemetery 4. Shou Qiu stelae 5. Shou Qiu mounds. *Credits* Images courtesy Google Earth, editing by the author

precisely under the Song, with the creation of the courtyards. Almost destroyed by fire, the temple was walled and reconstructed under the Ming in the version we can see today, which closely echoes architectural solutions adopted for the Forbidden City. The temple develops from south to north, and has its culmen in the Dacheng (Hall of Great Perfection), a huge pavilion reaching a height of 32 m supported by 28 richly decorated stone pillars, which serves as the main sacrificial hall. Annexed to the east of the temple is the Kong Family Mansion, the residence of the direct descendants of Confucius, who were dukes of the region and in charge of making special sacrifices in honor of their ancestor. The mansion, established by the Song and rebuilt under the Ming, is laid out with public quarters to the front—arranged in three subsequent halls—and residential quarters to the back, and houses a huge four-storey tower and an historical archive. Going on along the meridian lane and exiting the town to the north, one can follow a straight avenue decorated with a memorial arch up to, at some 1.5 km, the entrance to the third Confucian place, the so-called Forest of Confucius. It is a vast, evocative cemetery immersed in woods, where Confucius and many of his descendants have been buried over the millennia. Enclosed within a perimeter wall, it hosts the tombs of thousands of people, marked by memorial stones. Many tombs of the Dukes of Qufu, located to the northwest of the Tomb of their ancestor, have their own Spirit Paths with animals and guardian statues, built under the Ming. However, the Spirit Way of Confucius' own tomb, built under the Song, is especially notable for its enormous statues.

The sacred area pertaining to the Yellow Emperor, the Shou Qiu, is located some four kilometers from Qufu city center to the east. Shou Qiu is spoiled somewhat by modern buildings, and has been the object of vandalism in the past, but what remains is sufficient to see what a fascinating place it was. The Mausoleum develops in a south-to-north direction (azimuth around 7.5° east of north). At the northern end lies the "Hill of Longevity", where the Yellow Emperor was allegedly born. The tumulus (or natural hill, we do not know) was re-arranged by the Song in the twelfth century;

it is some nine meters high and resembles a small pyramid, covered with stone lintels. On its top sits a small statue pavilion. To the back of this pyramid stands a small mound considered to be the tomb of Shaohao, the son of the Yellow Emperor (the tomb of the Yellow Emperor himself is traditionally located in Huangling, Shaanxi). Besides covering the tumulus in stone, the Song also extended the axis of the Mausoleum by adding administrative and ceremonial buildings, today destroyed, and creating a spectacular entrance guarded by two immense, megalithic tortoise stelae. The stelae have been restored and today perfectly fulfill their ceremonial function, being more than sixteen meters tall. It is difficult to assess their weight precisely, but it cannot be much less than 200 tons (Fig. 5.11).

Yet another example of the intimate relationship of the Song rulers with the stones is of special interest for us, since it relates to the first Feng-Shui-adapted landscape historically documented in Chinese history: the *Genyue*. Apparently, the story of this place went as follows (Hargett 1988). Emperor Huizong (the last emperor of the Northern Song, ruling 1100–1126) was deprived of heirs. He asked advice to a famous diviner and the answer he received was:

> The earth at the northeastern corner of the Capital [Kaifeng] is in harmony with the Canopy [of Heaven] and the Chassis [of Earth]. Only its contours and conformations are somewhat low. If they were raised and heightened a little, the august heritors would then be abundant and ample.

In other words, Kaifeng needed a screen of mountains to the north east in order for the emperor to have sons. Apparently, the emperor took this advice very, very seriously, and decided to build a huge park complete with a chain of mountains

Fig. 5.11 Shou Qiu, Qufu. The eastern Tortoise stela

directly inside the capital. At that time, Kaifeng was one of the most populous cities in the world (population estimated at one and a half million people), and the Imperial Palace occupied the inner part and was surrounded by an enclosure wall. However, to make room for the park, substantial enlargement of the central area was effected to the north/north east, displacing the people and activities located in the expansion zone. The resulting park—which took the name of *Genyue*—was built over the course of four years (starting probably in 1117) by skilled artisans and stone cutters summoned from places renowned for such activities. The key feature of the park was, of course, the curtain of completely artificial mountains required by the prophecy. The maximum height was reached by a twin peak called (not by chance) by the emperor "Longevity Hills"; the actual height is debated: *cautious* estimates give it at around 28 m, which is already quite a respectable height, but other authors have estimated it to be four times as high. Be that as it may, the park must have been an astonishing masterpiece. An artificial waterfall descended from the artificial mountain to a goose pool, and the surrounding area was embellished with ponds, grottoes, and an infinite variety of stones, animals and plants sourced from all over the imperial territory. Huizong, in fact, established, with enormous expense, a nationwide network so that he could be sent beautiful stones, flowers, and animals. As is to be expected, some of the stones were gigantic: chronicles refer, for instance, to a block—placed at the entrance to the park and called "Divine Conveyance Rock"—which, if the translation of ancient measures is correct, must have been a megalith some fourteen meters long brought from Lake Tai (near Suzhou, some 800 km to the south east of Kaifeng). The transportation took many months and had to be made on a specially-adapted boat. All in all, Genyue was a new example of microcosm, similar to those we know of from Qin and from the Han Emperors, and also similar to other examples which still exist and will be encountered later in this book, like Chengde (Chap. 8). Conceived as miniature sacred landscapes, places like these were meant as replications of natural and/or man-made wonders of the Empire and therefore, of the Universe. The park was cardinally oriented and there exists textual evidence that at least some of its elements also replicated the geographical siting of their place of origin. Alas, the emperor's dream of longevity was not doomed to last. On January 9, 1127 the army of the Jin General Wolib entered Kaifeng. Most of the Genyue stones had already been broken up to produce throwing weapons during the siege; the trees and the woods of the palaces were burned in the next months to protect people from cold, and the animals slaughtered to feed them.

Last but not least, megalithic stones were also used for the monumental statues of the Spirit Paths of the Song Imperial Tombs. All these tombs are located in western Gongyi, Henan, about 120 km to the east of the capital Kaifeng, but near the ancestral home of the dynasty, the first Song Emperor' Taizu birthplace, Luoyang. After the retreat to Hangzhou, the emperors never lost the hope of reconquering the north, and for this reason only the mausoleums of the Northern Song rulers were built in monumental form. The Necropolis thus contains the mausoleums of the seven Northern Song Emperors, in addition to that of the father of the first emperor (who was posthumously proclaimed Emperor Xuanzu) and many satellites for empresses and court officials (Fig. 5.12).

Fig. 5.12 Mausoleums of the Northern Song dynasty, numbered in chronological order. 1. Xuanzu (Yongan) 2. Taizu (Yongchang) 3. Taizong (Yongxi) 4. Zhenzong (Yongding) 5. Renzong (Yongzhao) 6. Yingzong (Yonghou) 7. Shenzong (Yongyu) 8. Zhezong (Yongtai). E) Mausoleum of Empress Lu. *Credits* Images courtesy Google Earth, editing by the author

These mausoleums lack, of course, the grandeur of the Tang imperial projects, and are not very well-known to the general public; however, they are quite fascinating, adorned as they are with beautiful, and sometimes immense, statues. Unfortunately, the landscape they form is at high risk due to intense urbanization—for instance, the tomb of the second Song Emperor Taizong has almost been submerged by modern houses. Nonetheless, it is still possible to have an impression of how the royal Necropolis must have looked. It developed in a relatively restricted area (around 10 × 8 km) about 35 km from Luoyang city center, in the middle of a flatland which is bordered by relatively low hills on all sides except the west. The sacred "central" mountain of China, Mount Song, lies to the south-east at some 30 km distance. The mountain is visible as a familiar presence on the horizon, but it is only about 1500 m high and as such, is not terribly prominent (Fig. 5.13).

Each mausoleum develops in general south-north direction (for the precise orientations see below) and features a couple of earthen *que* pillars and a sacred way which ends at the burial mound. The mounds are notably smaller than those of the Han, and this is perhaps due to a rule—mentioned in historical sources—which stipulated that the tomb was not to be started in the lifetime of the emperor and had to be finished within seven months from his death. Apparently, this also led to a standardization of the layout of the monuments, which are very similar to each other. The spirit roads are short, the statues almost overcrowd and everything focuses the attention of the visitor on the mound at the end. However, the statues are finely carved and, in spite of the modesty of the projects, their size is impressive and their subjects quite original. They show a revived interest in the supernatural world, and at the same time reflect the political situation of the dynasty, obliged to confront powerful neighbors. The spirit world is thus represented, uniquely in the Song tombs, by the *jiaoduan*, a winged but burly "horse" credited with the ability of crossing mountains

Fig. 5.13 Yongtai, the tomb of Emperor Zhenzong. The Sacred Way viewed from the mound

and seas, carrying messages in many languages. Interestingly, just to show (as if it were necessary) the impressive cultural continuity of China, these animals are identical to those appearing on bronze vessels from a tomb at Anyang, Henan, dated to the Late Shang (1300 BC). Finely carved stone slabs representing phoenixes also feature. Real animals include rams, tigers and elephants guarded by "mahouts" from Southeast Asia or India. Officers include guards and eunuchs, but also foreign envoys depicted in fine detail with their own characteristic clothes and customs; for example, the statue of a Korean ambassador (more than three meters high) carries a bottle and wears an exquisitely carved, embossed metal hat (Fig. 5.14).

None of the Imperial Mausoleums has ever been excavated. Therefore, if we want to have an idea of how an Imperial Song underground tomb may look like and what it may contain, we must resort to burial sites from the same period, as we did with the Tang. Among excavated graves, a very important one is the tomb of an officer called Zhang Wenzao and of his wife, in the cemetery excavated at Xuanhua (Qingquan 2010). The tomb—closed in 1093 AD—is situated in a region which was then on the Song-Liao frontier and formally under Liao domination, but the owner was an ethnic Han. Actually, this tomb represents the epitome of Chinese funerary traditions, combined with the syncretist approach typical of the Neo-Confucian contemporary beliefs. The tomb is very small, almost miniaturized, but finely painted. It consists of a rectangular antechamber and a round burial chamber, connected by a narrow corridor. On the antechamber's wall, scenes of ordinary life are depicted: musicians and dancers on the west, and the preparation of tea—with children playing hide-and-seek—on the east. On the south wall, a lunette shows five grotesque figures

Fig. 5.14 Yongzhao, the tomb of Emperor Renzong. Auspicious *jiaoduan* on the Sacred Way

painted with pale blue ink. They are usually interpreted as "ghosts", but this would be, as far as I know, a one-off case in all Chinese funerary art, and a much more likely interpretation is that they are comic actors in a sacred drama involving the casting out of evil spirits (Hong 2013). The burial chamber seems to be built with the elaborate wooden ceilings typical of Chinese woodwork, but this is not so: the structures are made with bricks designed to imitate wood. Overall, everything in the tomb alludes to, and imitates, reality. For instance, false windows open into false wooden walls, confirming once again that things intended for the dead need not be completely functional. Real offerings in vessels appear together with miniature granaries, and we could go on: a painted servant lights up a real lamp on a real shelf protruding from the same wall on which she herself is painted, while another painted maid opens (or perhaps closes) a painted false door. Generally speaking, the activities of these characters (in this and other tombs of the period) seem to be related to the daily course of the sun: for instance, scenes appearing on the east and the west walls show people at their toilette when the sun rises and lighting a lamp when the sun sets. As for the main funerary and religious beliefs of the owners of the tomb, they were certainly Buddhist, since the bodies were cremated. However, bones and ashes were recomposed into kinds of "mannequins" buried in the coffins. In addition, a lunette in the funerary chamber shows three people engaged in a game on a checkerboard. It is easy to recognize that these three people are a Taoist priest, a Confucian Sage and a Buddhist Monk. Thus we can see that the owners of the tomb ultimately preferred not to take a definitive religious stance (Hsueh-Man 2005).

We now turn to the cognitive aspects of the landscape formed by the Song Necropolis. Paludan (1991) has proposed a connection of these monuments with a geomantic doctrine called "five sounds theory". She notes that, according to this theory, all sounds were divided into five groups, each one associated with a preferred direction. The sound of the Song family name had as auspicious direction, northwest-southeast, with mountains in the south and a slope towards the north, and Paludan finds these conditions satisfied by the Song tombs. However, the orientation of the Song tombs is not northwestward, it is actually slightly to the east of north (see below), and it is not directed to the mountain peaks; the only verifiable thing is that the spirit roads slope downwards slightly, but this looks more like a condition forced by topography. Therefore, although it is known that in the Song period, music played an important part in the construction of cultural identity (as shown, for instance, by the magnificent sculptures of musicians in the tomb of Wang Jian in Chengdu, built around 918 AD), the solution should be sought elsewhere.

In my view, to understand the landscape of the Song mausoleums, the first thing to be noticed is that they were placed according to a definite rule—similar to, but different from, the Zhaomu (Magli 2018). This rule requires the mausoleums to be arranged in pairs, with the second element always positioned slightly to the northwest of the previous one—so perhaps this one choice was really due to the "five sounds" theory, since in this way the mausoleum of the ancestor would be propitiously placed to the southeast. Interestingly, the only exception is the tomb of Emperor Zhenzong, which seems quite isolated. However, to the northwest of it there is the only satellite Song mausoleum of any significance. It is the tomb of Empress Liu, Zhenzong's wife. She actually served as regent of China during the illness of her husband (1020–1022 AD), and later, while her son, Emperor Renzong, was still a minor. So it appears that the rule was also at play in this case.

Finally, as in the case of the Tang mausoleums, to study the orientations it is natural to measure the spirit paths (rather than the mounds, which are mostly now dilapidated). These orientations are given in Table A.3: seven out of seven fall between 0° and 5° that is, the Spirit Roads point slightly to the east of north. The deviations seem to be too high to be attributed to errors in determining true north. A deviation of ~5° is, on the other hand, consistent with the fact that, during the Song Dynasty, the north celestial pole was slowly completing its approach to Polaris, the elongation being ~6° in 960 AD reducing to ~5° 30′ during the 150 years spanned by the seven emperors. It is thus conceivable that these monuments were oriented—like the Tang ones—to the maximal (eastern in this case) elongation of Polaris. As far as a possible magnetic orientation is concerned, it appears that no correlation exists between the orientations and the estimated values of magnetic declinations.

5.3   From Mountains to Pagodas

The *Tanguts* were a tribe originating in the Tibetan Qinghai region, which migrated to the east during the seventh century, settling in northern Shaanxi during the Tang rule. On the collapse of the Tang Dynasty, they managed to gain sufficient territories and independence to enable them to build an independent state in 1038, when Li Yuanhao declared himself the first Emperor of the Dynasty we call Western Xia, with capital Xingqing (modern Yinchuan). The Song were unable to overthrow the Xia state, which continued to exist for two centuries, extending essentially in what are now the northwestern Chinese provinces.

Our knowledge of the Tanguts culture and society is very limited; we know that Tantric Buddhism was practiced and that the emperors ordered the creation of a Tangut writing script and the translation of the Classics into it, but original documents are almost completely lost. The reason lies in the terrible threat that started to materialize at China's borders at the end of the twelfth century. Around 1190, in fact, a Mongol tribal head called Temujin managed to reunite the Mongol tribes under his leadership. Having being proclaimed Supreme Ruler or *Genghis Khan*, in 1205 he turned his attention to the world outside, and, in particular, to the Western Xia who had granted asylum to some of his opponents. This started a long period of wars which ended definitively in September 1227, when the last Western Xia Emperor surrendered and the Mongol army literally annihilated the capital. As a consequence, few monuments of the period survive; among them, a very peculiar one is the so-called 108 stupa monument, located on a hillside directly on the western bank of the Yellow River, south of the capital. It is composed by 108 stupa of sun-dried mud bricks, arranged in rows disposed in a triangular formation which narrows with height, from 19 stupa on the first row to the uppermost single one. The monument (today heavily restored) is oriented to winter solstice sunrise and bears a certain resemblance with the famous Javanese temple of Borobudur, constructed slightly before (Magli 2020).

Among the devastation wrought by Genghis Khan's troops, there was considerable destruction of the imperial tombs. In spite of this, however, much still remains of these curious monuments. The Imperial Xia tombs are located in the western outskirts of Yinchuan at the foot of the Helan Mountains. The Necropolis seems to have been laid out from south to north and the royal tombs today are numbered accordingly (Fig. 5.15).

There are hundreds of other tombs, but there can be little doubt which are the royal ones—which are nine—because they are the only one that feature the strange mound we shall shortly discuss. It is difficult to identify which emperors are the possessors of each of these tombs however: originally, they had inscribed stelae, but these stelae were all destroyed by the Mongols and only fragments have been recovered. On the basis of these fragments, it has been possible only to ascertain that the owner of Tomb 7 is the fifth Xia Emperor, Renzong. This may seem odd, but it makes sense, because it is known that the first emperor upgraded both his father and his grandfather to imperial state posthumously, and buried them in the royal Necropolis. Their tombs should therefore be those numbered 1 and 2, which are located at the southernmost limit of the Necropolis (and oriented differently, see below). If the development from south to north is accepted, then the evolution of the Necropolis seems also in some agreement with the "northwest" rule used by the Song. However, the Xia had in total ten emperors, so the tombs of the last three emperors are missing (Steinhard 1993).

Fig. 5.15 The 9 imperial tombs of the Western Xia dynasty; U denotes the unfinished project. North to the right. *Credits* Images courtesy Google Earth, editing by the author

Most probably, the tomb of the very last emperor (put to death by the Mongols) was never built. It is accepted that neither those of the eighth and ninth were built, but if this might be reasonable for the ninth, who reigned for only three years in the throes of a terrible war, it looks less convincing for the eighth, Shenzong, who reigned for twelve years. Actually, despite the fact that the northernmost zone of the Necropolis is much disturbed (it appears to have been overbuilt in recent times, and then cleared again) it seems to me that an unfinished royal project can be identified slightly north-east of Tombs 8 and 9, a thing that, if verified, would solve at least this riddle. Others, however, remain unsolved: one only has visit the necropolis to appreciate this (Fig. 5.16).

The plan of a Xia tomb is somewhat similar to that of the Northern Song tombs: a rectangular enclosure where from south to north we encounter a pair of decorated *que* followed by coupled stelae pavilions and by a sacred way (admittedly, some statues—or perhaps stelae bases—are quite strange, showing masks glowering with grinding teeth). The tomb, at the end of the sacred way, is located under a very peculiar mound: a huge layered structure of rammed earth, which was originally faced with bricks and octagonal in plan. The mounds are usually slightly displaced to the northwest with respect to the main axis (for unknown reasons), and today—frankly speaking—they recall nothing so much as large termite hills, the tallest one reaching the respectable height of 23 m. However, the recovery of tiles and of other elements has led archaeologists to believe that they originally resembled a decorated Buddhist Pagoda of octagonal base, featuring six or seven levels. These buildings are perhaps those represented in a Buddhist cave of the period located along the Yulin River near Anxi, where Buddha is shown as enthroned on a stepped mound.

As regards the Necropolis as a whole, a few observations can be made. First of all, there is no evidence whatsoever that the tombs were placed following an astronomical pattern resembling the Big Dipper, as sometimes stated in the literature (Wang 2016). As far as the orientations are concerned, the first two tombs, probably

Fig. 5.16 Imperial tombs of the Western Xia dynasty, tomb 3

for the ancestors of the dynasty, are quite precisely oriented to the cardinal points. All the others are considerably skewed to the west of north (Table A.4). There is too much uncertainty about the owners of the tombs to put them in chronological order and construct a reasonable correlation curve with the magnetic declination at the site, but a magnetic orientation can, in any case be ruled out. Indeed, here, magnetic declination between 1038 and 1227 AD changed from −3.6 to +0.74 and we are therefore very far from the observed deviations. A simple explanation for the orientations does, however, exist and it is to be found in the landscape: the tombs were meant to form a harmonious view with the mountains in the background, in a way similar to what has been shown to happen to the tombs of the Tibetan kings (Romain 2019). Indeed, although in the present case no specific orientation to mountain peaks appears, the sides of the enclosures are parallel with the hill's veins.

Now it is time for us to get accustomed to such spectacular views, for we are going to be encountering many of them, and much more complex indeed.

Chapter 6
A New Splendor

6.1 The Advent of the Ming

The Song Dynasty was not destined to endure much longer than the Western Xia. In fact, in 1271, Genghis' grandson Kublai Khan, on the point of invading the Yangtze basin, declared that a new Dynasty was ruling China: the Yuan (Kublai Khan is the Chinese emperor, with whom the famous Italian explorer Marco Polo probably met in 1275). The Mongol conquest was completed in 1279. Kublai settled his capital in Khanbaliq (the future Beijing), establishing also a summer capital at Xanadu in the north. The Mongols adopted the Chinese syncretist worldview, and were probably fascinated by Feng Shui principles in architecture too (Romain 2017). The dominant Mongol class preserved, however, traditional nomadic customs and, in particular, the emperors were not buried in mausoleums but in the earth of homeland.

The Mongol-led rule was not to last long. In the second half of the fourteenth century, a popular uprising against foreign domination, combined with natural disasters and poor governance, finally led to rebellion. Among the rebels, there emerged the figure of Zhu Yuanzhang, born into a poor family of farmers in present-day Fengyang, Anhui Province. In 1356, Zhu and his army conquered Nanjing, which was to become the first capital of the new rulers, the Ming Dynasty, proclaimed in 1368 on the complete reconquest of the country. Zhu took the reign title of Hongwu (meaning something like "eminently martial"). As mentioned in Sect. 4.3, starting from the reign of Emperor Wu of Han, each emperor assumed a reign title or *era name* used to count the years; these names were not fixed, however, and could be changed on occasions. With the Ming (and later the Qing), on the other hand, each emperor (with only one exception) had a single reign title, which are therefore used today to refer to them.

The reign of the Ming lasted up to 1644 and marked a period of stability for the Chinese empire. From the very beginning, Hongwu based the Ming rule on an extensive revival of the Tang and Song traditions and values: in particular, ancestor worship was reinstated. He decided to honor his own ancestors giving them a posthumous imperial status and for that reason, ordered the construction of an imperial mausoleum for his parents (Huangling, at Fengyang), as well as the construction of a second imperial tomb (Zuling) for his ancestors on the banks of Lake Hongse, Jiangsu. This monument was flooded in the seventeenth century, but it was miraculously rediscovered in the final decades of the last century, with all the fine details of the statuary preserved intact by the mud. Hongwu also played upon the traditional semi-divine image of the ruler and restored in full the idea of the Mandate of Heaven, which his successor Yongle would allow to inform his building program when he moved the capital to Beijing, as we shall see.

The revival of lost traditions and the inspiring ideas which were to characterize Ming architecture and landscape are especially apparent in the tomb of Hongwu himself, the Xiaoling Mausoleum. Xiaoling is an undisputed masterpiece, which set the benchmark for the tombs of all the subsequent Ming rulers. The complex is located at southern foot of Purple Mountain, which—despite its name—is a rather low-lying hill located just east of Nanjing. Works at Xiaoling lasted up to the first decade of the fifteenth century, during the reign of Hongwu's successor, and involved a considerable amount of resources and manpower.

Entrance to the complex is through a "dismounting archway" (the limit beyond which riding horses was not allowed) and a monumental gate followed by a huge square pavilion. Inside the pavilion, one encounters one the most astonishing megalithic artifacts ever made on earth, the Shengong stela. It is a tortoise stela, that is, a monolith held up by a tortoise statue and crowned with a capital. The tortoise is more than 5 m long, 2.5 m wide and 2.8 m tall. Its weight is difficult to estimate accurately but it cannot be less than 80 tons. The stela on the back of the tortoise reaches 9 m and bears an inscription on the merits of Hongwu signed by the successor, who erected it. Even though it is of its staggering size, as we shall shortly see, it is probable that this monument was originally planned to be as much as ten times larger (yet another giant tortoise stela has been recently recovered in pieces near the pavilion; its builder remain unknown) (Fig. 6.1).

The Sacred Road starts after the stela pavilion. It is divided into two parts, both flanked by monolithic statues. The first part is oriented 22° north of west and is lined with pairs of animals: real animals—elephants, horses, and camels—depicted life-size, and mythical beasts, like the *qilin*, here represented as a beast the size of a oxen with dragon's head and horns, traditionally associated with auspiciousness and good governance. Then the road bends, tracing first a sort of elbow shape and then aligning with the final part of the mausoleum, which is oriented very precisely to true north. It is usually argued that the elbow was made to avoid (as a sign of respect) the mound tomb of a local king of the third century AD, Sun Quan (actually I was unable to identify this tomb on site, apart from the presence of a modern statue). Be that as it may, the main issue is that the Spirit Road macroscopically bends, and this also has a symbolical value. Indeed, according to Chinese tradition, evil spirits are confused by angles and gates, and this idea was clearly taken on board by the Ming architects, who started to introduce along the axes of the tombs perplexing (for us) gate "screens" (that is, not opening into walls but self-standing) and indeed, bends (Fig. 6.2).

Fig. 6.1 Nanjing, Xiaoling tomb. The Shengong stela

The second section of the road has a different subject matter: it is lined first by pillars and then by human figures of imperial officers. There is no attempt to depict real-life people: their expressions are solemn, with a blank look, and the sights are lost, similar to those of the terracotta army. Overall, the Xiaoling Spirit Road (and that of the thirteen tombs valley we shall visit later) is quite different from those of the Northern Song tombs: foreign ambassadors and elephants with grooms, for instance, are not present and, generally speaking, the symbolic and ceremonial aspect is much more strongly accentuated. The closest parallel is thus with the Tang Spirit Roads; a more relaxed style closer to the Song one was, however, to remain in use in the statuary for local kings and nobles—one beautiful example being the noble's Ming tombs in Guilin; another the tomb of a Ming Prince of Xi'an which is located about 1.5 km south-west of Emperor Xuan mound.

Fig. 6.2 Nanjing, Xiaoling tomb. Stone elephant on the Sacred Way

At the road's end, a gateway leads us to the next section of the complex: a rectangular courtyard. It is sometimes claimed that the design of the Ming tombs expresses the Chinese idea that the Earth is square (reflected in the shape of the courtyard) as opposed to Heaven, which is round (reflected in the round shape of the mound). This is a rather naive interpretation: on one hand, the "square" architecture of the courtyards—which we first encounter at Xiaoling and will meet again in the next tombs—is more than mere "architecture for the living": it is almost a copy of what can be seen in the imperial palaces. In these tomb buildings, the emperor would come regularly to celebrate rites in honor of his ancestors. On the other hand, the tomb itself—although covered by a round mound—is an underground palace, rectangular in plan.

After a series of archways, bridges, and buildings dedicated to the worship of the deceased (today in various states of conservation), we finally reach the tomb proper. In front of the visitor appears the *Ming Lou* or Soul Tower: a very tall building (the front wall is about 20 m tall) including a gateway and a pavilion on top; inside the gateway, a monumental staircase ascends. I will always remember the first time I saw this apparently nonsensical monument: it is so big that it is difficult to capture it in one photograph. In later Soul Towers, the pavilion on top would accommodate a huge stela; here it is empty, but the meaning is clear. Indeed, when you reach the top and look north, you will see the huge mass of the tomb's earthen mound; when you

Fig. 6.3 Nanjing, Xiaoling tomb. The Soul tower

look south, you will see the tomb complex extending towards Nanjing, ideally with the same view of the death emperor who overlooks the country from its northern, imperishable seat (Fig. 6.3).

6.2 Three Stones for a King

The appointed heir of Emperor Hongwu was Zhu Yunwen, who ascended to the throne as the Jianwen Emperor. However, his Uncle Zhu Di, Prince of Beijing, rebelled against him, conquering Nanjing in 1402 and proclaiming himself Emperor with the name Yongle. The historical figure of Yongle is controversial. On one hand, his ferocious persecution of adversaries and of—real or presumed—traitors is legendary, and some of his ruthless, large-scale enterprises involved the infliction of atrocious suffering on many people; on the other, in his hands the economy of the empire flourished. Among his projects, he ordered the extensive renovation and restoration of the Grand Canal, an artificial river linking the Yellow River and the Yangtze River (some sections of this astonishing engineering feat had already been built by the fifth century BC, but the work was completed under the Sui, resulting in a total length of more than 1750 km). Fascinated by geography and sea voyages, by 1403 Yongle had tasked three fleets with traveling up to Java and India to proclaim his accession, seeking acknowledgment of his power. Among the kingdoms to pay homage to him was Boni, whose king made an official visit to Yongle, reaching Nanjing in August 1408. However, he suddenly fell ill and died in the same year, having expressed the wish to be buried in China. Yongle ordered this wish to be fulfilled and had a sumptuous tomb constructed for him. The tomb, located in the southern suburbs of Nanjing, still exists today and features a beautiful Spirit Way in the style of Ming nobility.

Among the political missions launched by the emperor to increase trade and international awareness of his power, the most famous are those of the Muslim eunuch Admiral Zheng He. Zheng He (born Ma He, the honorific title Zheng was conferred on him by the Emperor) was the son of a Mongol officer; captured very young and castrated, he was put at the service of Prince Zhu Di, the future Yongle Emperor, in Beijing. He acquired the trust of the Prince and assisted him in his accession to the throne, serving as general in the conquest of Nanjing. Once the empire was settled, Yongle appointed Zheng Admiral and Chief Envoy, sending him on several voyages as chief of the enormous royal fleets and armies. Zheng visited Brunei, Java, Thailand and India. He also circumnavigated the Indian subcontinent, reaching the Red Sea and traveling along the eastern coast of Africa. At each stage, he presented gifts and exchanged goods: Chinese items in gold and porcelain were exchanged for the most disparate things, including a living giraffe which, once in China, was of course considered a very auspicious exemplar of a qilin.

The wealth of the state finances also allowed Yongle to carry out an impressive building program. In particular, it was under Yongle that the Ming start to consider the idea of constructing a stone wall at the northern boundaries. This, of course, is the monumental wonder which was to become famous as the Great Wall, although it is never called such in original Chinese documents (Waldron 1990). The work lasted for many decades, with a substantial impetus in the first half of the sixteenth century; for instance, the (today much visited) Badaling stretch was built in 1505. The construction of boundary walls was certainly nothing new for China (the first emperor had already built thousands of kilometers of walls of rammed earth) but there is no doubt that the Ming took this idea to quite another level. The Ming wall is some 8850 km in length, of which ¾ was physically constructed (the other sections take advantage of natural barriers). Anyone who has visited a few sections of this masterpiece, which runs uninterrupted along the crests of the hills, is amazed at the sheer beauty of the wall, which belies both its defensive nature and the backbreaking efforts its building entailed. This is, of course, not the place for such a discussion, but the symbolic and cognitive role of the Great Wall (especially in the stretches directly protecting the capital) has not, in my view, been sufficiently investigated. It seems obvious that—except for specific sections and garrisons built with clear strategic aims—the wall itself can function as a deterrent structure as well as an efficient instrument of early warning in case of invasions, but cannot (and in fact, was not) be an effective method of in situ defense. So some other factors, including, above all idea of explicitly asserting the boundary between the civilized world and the barbarians, must have played a significant role in its construction. In a sense, then, the Great Wall is the boundary of a sacred space, and as such was—hopefully of course—inviolable.

The construction program of Yongle reached its apex when he decided to move the capital to his former principate, Beijing. But, before transferring to Beijing, Yongle was apparently responsible for the most insane architectural project ever attempted on the planet, a project we are now going to visit: Yangshan. Yangshan is a limestone quarry located some 10 km to the east of Purple Mountain. Huge, regular cuts in the rock veins show that blocks of stone were quarried here for many centuries, but the reason why this place is so very special is because of three blocks which still remain on site. The quarry is at the foot of a hill, and the first of these blocks is visible as one ascends. It was detached on three sides; under the block many regular cavities (large enough to accommodate a group of kneeling men) can be seen. These cavities were clearly meant to be used at the moment of definitive extraction, by hammering the stone diaphragms remaining to support the block from below. The block itself has been carefully crafted to make a smooth surface on the detached sides and has the shape of a parallelepiped with slightly rounded angles at the front. And now, its size. It is about $30 \times 13 \times 16$ m (yes), giving a total of more than 6200 m^3. If we take a very cautious estimate of the weight of the Yangshan quality of stone at 2.5 metric tons per cubic meter, the block comes out at about 15,500 tons (Fig. 6.4).

Before making any comment, let us proceed. The path rises slightly up to an esplanade, where a second block of stone appears. This one has been cut free on all the

Fig. 6.4 Nanjing, Yangshan quarry. Block 1

Fig. 6.5 Nanjing, Yangshan quarry. Front view of Block 2. Notice the huge cavities at the basis, made for detaching the block, and the notches on the surface (author is shown for comparison)

sides, and only the base (deeply excavated with cavities similar to the ones described above) remains anchored to the rock. The shape is again that of a parallelepiped, which, however, has been rounded at all corners. Huge protuberances have been deliberately left here and there, giving the impression of decorations. The size is around $10 \times 20 \times 8.5$, so that the block has a volume of more than 1700 m^3, weighing not less than 4200 tons (Fig. 6.5).

To the left of this object, runs a wall of stone which has been perfectly leveled on the surface. Under the wall runs a huge, almost continuous gallery, the height of a slightly kneeling person, almost perfectly carved. It takes a while to understand that this gallery is nothing but, again, the detaching cavity of a third block of stone, as long as 50 m, 11 m wide and more than 4 m thick, for a total of some 2200 m^3 or 5500 metric tons (Fig. 6.6).

We are now (more or less) ready to understand that the project I have just described is that of a tortoise stela: the first block was to be sculpted as a turtle so enormous that the head would reach up to 16 m, block 2 was the capital, to be fitted on top of the stela body (block 3), giving a final height of over 70 m. There is no doubt that even if only the smallest of such blocks had been moved just 1 cm from the quarry, it would have been the most astonishing feat ever achieved in the history of engineering before electricity, and indeed a highly memorable one in the history of technology generally. To have an idea about weight limits today, we might recall that

Fig. 6.6 Nanjing, Yangshan quarry. Rear view of Blocks 2 (left) and 3 (right)

the most powerful movable crane barely reaches a load limit of 1200 tons, and that greater weights (extremely rare) must be moved with girdle cranes sliding on fixed rails.

In short, the enterprise of moving the stones from Yangshan was, not to put too fine a point upon it, impossible. They were conceived for Xiaoling, where Yongle wanted to legitimize his own power (which, as we have seen, was something of a usurpation), constructing for his father a memorial stela never been seen before. In fact, he almost achieved that anyway, since the Shengong stela which was successfully put in place is comparable only to the Song Stelae erected in Qufu, but the Yangshan project in itself was crazy and had to be abandoned. We do not know the details of such an abandonment: I cannot believe—as is sometimes stated—that the architects did not know that moving the stones was impossible, so perhaps they did not have the nerve to tell the Emperor the truth until the fateful day for moving the stones was imminent. The blocks were left behind, and the project of the stela was scaled down to the—still gigantic—size we can see still today in Xiaoling.

In the meantime, the Emperor's interest in the project had probably begun to wane, as he started to work to his own, new capital.

6.3   The Cosmic Capital

We do not know why Beijing became the capital of the Ming Empire; it is sometimes suggested that it was for strategical reasons—in order for the emperor to be nearer to the turbulent northern border—but if that was so (and I do not believe it), it did not prove to be a good idea at all.

Rather, Beijing was the principate from where Yongle contrived his ascent to power, and this must have played a significant part. Whatever the motive, for his triumphal return as Emperor, Yongle set in action a massive building program which, starting in 1407, was virtually complete 13 years later (Fig. 6.7).

The Ming city walls of Beijing are no longer visible as they were almost completely removed in 1950, but they essentially corresponded to the circuit of the modern Beijing's third ring road. The program involved the complete rebuilding of the former Yuan city with the objective of tangibly expressing that Beijing was the place of residence of the person in charge of the Mandate of Heaven: this place took the name *Zijincheng*. *Zijincheng* literally means *Purple Forbidden City*, the adjective "purple" being an explicit reference to the Purple Enclosure, the region of the sky near the north celestial pole whose stars were identified with the emperor and his court. From the very beginning, therefore, Beijing was a "cosmic capital" and its core, the place that today we simply call the Forbidden City, was identified with the earthly counterpart of the most important region of the heavens. One should not, however, think of a terrestrial map of the constellations on earth or of some other esoteric meaning for this place, which—being on earth—was, rather, planned in accordance with strictly terrestrial rules, that is, Feng Shui (Fig. 6.8).

The Forbidden City is a rectangular, walled enclosure (about 1000 m north-south by 750 m east-west) surrounded by a wide moat. There are four towers at the four

Fig. 6.7 The main monuments of Ming Beijing mentioned in the text: (1) Forbidden City—Hall of Supreme Harmony (2) Fengshan Hill (3) West Palace (4) Altar of Grain (5) Temple of Earth (6) Temple of the Sun (7) Temple of Heaven (8) Temple of the Moon. *Credits* Images courtesy Google Earth, editing by the author

Fig. 6.8 Beijing, The Forbidden City from Fengshan hill

corners, and four gates, one per side. The most important gates are the Meridian Gate—with its protruding wings—to the south, which allowed (and still does allow) the public to enter, and the northern gate (Gate of Divine Might), which connects the imperial apartments area with Jingshan Park. The plan of the Forbidden city is characterized by extremely rigorous geometrical layout. Generally speaking, it is divided south-north into a public (southern) part and a private (northern) part, and it is divided east-west in three sections, the central one being reserved for the emperor and crossed by an ideal imperial path reserved for him. Entering through the Meridian Gate, the visitor enters a vast square. A meandering, canalized river (the "Golden Water" River) flows across the square and is spanned by five bridges (Fig. 6.9).

To the north stands the imposing mass of the Gate of Supreme Harmony. Behind that, a second esplanade prepares for the focus of the complex, a succession of three halls, or pavilions, built on a raised terrace: the Hall of Supreme Harmony, the Hall of Central Harmony, and the Hall of Preserving Harmony. The central ascending ramps to the halls are part of the imperial route and were reserved for the emperor's baldachin; they are made of huge blocks of stone carved with fine reliefs of dragons and phoenixes, symbols of the royal couple. In particular, the northernmost ramp leading to the Hall of Preserving Harmony was carved from a huge block of stone almost 17 m long; the southernmost ramp is quite similar but made of two blocks. The interiors of the halls were devoted to imperial duties and each one hosts a throne. In particular, the Hall of Supreme Harmony is the largest and most symbolic building in the Forbidden City; its wooden columns reach a height of more than 20 m. It was used for court and ceremonial occasions, and it is the largest wooden structure existing in China. Its style inspired deeply Ming official architecture throughout the country. The second hall or Hall of Central Harmony is a square building used as a transitional place for the Emperor to rest during ceremonies. The third hall or Hall of Preserving Harmony is famous mostly because it was the place where the final stage of the laborious process of Imperial State Examination took place (Fig. 6.10).

Fig. 6.9   Beijing, Forbidden City. The Golden Water river and the south entrance hall

Fig. 6.10   Beijing, Forbidden City. The hall of Supreme Harmony with the summit of Fengshan hill visible on the far left

The private part of the Forbidden City, or inner court, where the court lived, is located in the northern section and ends in a small private garden. This section was also designed following a rigorous, geometrical plan and its central part is a sort of mirror image of the central part of the public section, based as it is on the succession of three halls, called the Palace of Heavenly Purity (the Emperor's residence), the Hall of Union, and the Palace of Earthly Tranquility (Empress's residence). A very famous, strikingly distinctive feature of the Forbidden City is the edges of the roofs of these buildings, which are embellished by statuettes. In a row, one can always see a man "riding" a bird and a dragon; other mythical beasts may feature, and the number of statuettes increases with the importance of the building.

The Forbidden City is incontestably one of the most astonishing architectural feats of mankind. I have only used a few lines to describe it, but I believe that not even a complete book would be able to convey the emotion of seeing the complex from the top of Jingshan hill (the artificial hill to the north, see below), when the red walls emerge from the morning mist of Beijing. The Forbidden City was conceived as "a Feng Shui landscape where there is no Feng Shui landscape". First of all, orientation is to the cardinal points. The axis deviates about 2° west of north; whether this deviation was due to magnetic orientation or not is difficult to ascertain, as the magnetic declination at the date was around 1° west of north (it is sometimes noticed that the northern prolongation of this axis passes closely to Xanadu, the summer capital of the Mongols, but—even admitting that the axis of Ming Beijing preserved that of the Mongol capital—this "alignment" is too long to be effective). Besides orientation, the planners had to cope with the problem that Beijing lies on a very flat land. Indeed, it is true that the city has mountains only to the north, the east and the west and a river to the south, but all this is tens of kilometers away. So, in spite of the efforts of later geomancers to exalt the presumed, auspicious position of the capital (Meyer 1978, 1991) it is very clear that the Ming architects who planned the Forbidden City decided to adjust the two main features of any auspicious landscape (mountain to the north, river to the south) artificially. Indeed, they *added* these two things locally. The Golden Water river was canalized to flow across the southern esplanade, and the Jingshan Hill (more than 50 m high) was simply built from scratch—probably with the material excavated from the moat's pits—to serve as mountain of the north. This is really a curious, twofold feature of the final (Ming and later) Feng Shui conception of the landscape. A landscape is indeed seen as a modifiable being, which can be damaged, but also ameliorated, with consequent variations in the flow and the accumulation of Qi.

The Ming planning of the Forbidden City influenced the overall urban layout of Beijing, as the southern prolongation of the main axis through Tiananmen gate (and today Tiananmen square) became the central axis of the town. The result is a city which, due to the series of walls and gates and to the interplay between open and closed, "folds and unfolds" as a papyrus scroll (Zhu 2003). In particular, Yongle constructed a series of sacred complexes devoted to the explicit celebration of the role of the Emperor as intermediary with the Celestial Gods and to the renewal of the Mandate of Heaven. First of all, near the south west corner of the Forbidden City (today Zhongshan Park), the Emperor built the Altar of Land and Grain. This

Fig. 6.11 Beijing, Temple of Heaven, Hall of Prayer for Good Harvests

is a three-level square terrace of white marble, used for making sacrifices to the Earth's gods. On the top, a basin contains soil of five colors: green (east), red (south), white (west), black (north) and finally yellow in the middle. On the opposite corner (south east) of the Forbidden City, another sacred building was located: the Imperial Ancestral Temple, where Ming ancestors were venerated. Proceeding further south after Tiananmen, and parallel to the main axis, we meet yet another huge complex originally built by Yongle: the *Temple of Heaven* (Fig. 6.11).

Today incorporated in a public park, it originally included two main buildings. First (from north to south) is the Hall of Prayer for Good Harvests, a high, circular wooden hall, built on a three-level marble terrace. Constructed in 1420, it is a master-piece of Chinese woodwork: the vault is able reach such a height due to the presence of three eaves. The second complex, linked to the first by a processional walkway, is the Imperial Vault of Heaven, similar to the Hall of Prayer for Good Harvests and again built on a white marble platform, but smaller. It is enclosed by a curved wall, traditionally called the Echo Wall because of its (true or rather presumed) acoustic characteristics. The third main building of the Temple of Heaven, the Circular Mound Altar, was added one century later by Emperor Jiajing, who was the responsible for the definitive "cosmization" of the Ming capital (Fig. 6.12).

Jiajing was a frenetic builder. His first projects stemmed from the fact that he was not the son of his predecessor Zhengde, who died without direct heirs. Once he ascended to the throne, he refused to be formally (posthumously) adopted, as tradition would have imposed. Rather, he elevated his parents directly to imperial rank and accordingly ordered the construction of an imperial mausoleum for them in their homeland, Zhongxiang (see Sect. 7.2 for a description of this monument). In Beijing, Jiajing embarked on a very complex building program. First of all, as mentioned, he completed the Temple of Heaven with the addition of a third, spectacular building along the main ceremonial axis. Called the Circular Mound Altar and built around 1530, it is an open-air, three-layered, circular platform of white marble decorated

Fig. 6.12 Beijing, Temple of Heaven, The main axis of the complex viewed from the Altar of Earth

with stone dragons. Jiajing also added a curious stone group to the park situated east of the Hall of Prayer for Good Harvest: seven massive stones carved with the traditional Chinese form of idealized, conical mountains, said to represent the peaks of Mount Tai, the sacred mountain of the east. The extension of the Temple of Heaven was, however, only a part of the ambitious building program of Jiajing, aimed at the completion the "cosmization" of Beijing. In fact, in addition to the Circular Mound Altar which, being in axis with the processional way of the Temple of Heaven, is located to the south of the Forbidden City, Jiajing also built three other temples, placed at the other three cardinal points: the so-called Temple of the Sun in the east, the Temple of the Earth in the north, and the Temple of the Moon in the west. The Temple of the Earth is today set in a beautiful public park, Ditan, and it is a vast, open air altar. Being dedicated to the Earth Gods, it is square in plan in line with the Chinese association of Earth with the square shape. The Temple of the Sun is today located in Ritan Park, while what remains of the temple of the Moon can be seen in the Yuetan Park. To my surprise, I found that the design of these temples was strongly interconnected, as very clear topographical relationships link them with each other. Indeed:

- the center of the altar of the temple of Earth lies almost on the same meridian of the center of the circular mound altar (distance 8.4 km, error less than 1.5°)

- the parallel of the center of the Altar of the Temple of the Sun crosses the heart of the Forbidden City and was probably meant to connect with the center of the Hall of Supreme Harmony, which is missed by less than 150 m (distance 4 km, error in degrees less than 0.5°)
- the parallel of the center of the Altar of the Temple of the Moon passes through the Hall of Supreme Harmony with no discernible error (distance 3.8 km)

The last two alignments are not astronomical, as the sun would not have been visible rising/setting along the east-west line with the town in between, but are sufficiently short to be measurable with the help of high poles, starting from a measurement of the cardinal directions that was certainly astronomical and probably based on the sun (not magnetic). The temples are associated with the yearly cycle and evidently the Emperor made sacrifices in all these places on a yearly basis, in coincidence with the autumnal equinox (Temple of the Moon), the summer solstice (Temple of the Earth), the spring equinox (Temple of the Sun), and the winter solstice (Circular Mound Altar) respectively. As a matter of fact, to the southwest of the Circular Mound Altar stands a very high pole which was used—together with two others nearby of which only the foundations remain—for hanging a lantern on. I have verified that the position of this complex in relation to the center of the altar mound could be useful as a signpost to check, at sunset, the correct date of the solstice.

As far as the orientation of each of these structures is concerned, that of the Temple of Heaven is consistent with those of the Forbidden City and the Altar of Soil, which are contemporary, at about 2.5° west, The orientations of the other three temples are consistent with each other (Temple of Sun 4.5° west, Temple of Earth 5° west, Temple of Moon 4.5° west) and it is possible that these orientations were magnetic, since they are qualitatively in accordance with the behavior of the magnetic declination at Beijing at the time of construction.

The final "cosmization" of Beijing was underway when, on the night of November 27, 1542, a group of palace maids attempted to strangle Emperor Jiajing. The plot did not come off, but the life of Jiajing changed forever. Terrified by the idea of continuing to living in the Forbidden City, he decided to move to the palace that had been the residence of Yongle at the time he was the Beijing Prince. This palace, called Yongshou, is located in a vast park that extends along the west side of the Forbidden City (West Park is today a Government residence and is closed to general public). This move triggered yet another major step in the Emperor' building program (Wang 2009). Jiajing was a devoted Taoist; after that the attempted assassination failed, he retired from state affairs and turned his attention to alchemy and elixirs, in an attempt to attain immortality. The West Park (founded by the Yuan and renovated under Yongle) was already a fascinating place, with three interconnected lakes, but the Emperor introduced there stone mounds, caves and pavilions with the aim of making it resemble an "immortal land": a place suited for Taoist spiritual transformations and rituals, adorned with magnificent buildings and populated by exotic animals.

So we have yet another microcosm created for an emperor's delectation, but—in this case—it replicated an imaginary world.

Chapter 7
A Beautiful Valley

7.1 From Nanjing to Shisanling

When Yongle moved from Nanjing to Beijing, he was also compelled to choose a new site for his tomb, not too distant from the new capital. The place chosen is called Shisanling, and is located in Changping District, 42 km to the northwest of the Forbidden City. With the construction of his tomb there, named Changling, Yongle inaugurated a necropolis which was subsequently used by all of his successors, a place of incredible beauty and peace today known as the Thirteen Tombs of the Ming Dynasty. All these tomb projects were inserted harmoniously into the already existing ones. To have an understanding of the place, we must imagine it as it was when the Emperor's architects first saw it, and compare the Yongle project with that, just completed, of the emperor's father in Nanjing (Paludan 1981) (Fig. 7.1).

The valley is a verdant plateau, roughly oriented south-north, with the Mountains of Tianshou to the north and smooth hills to the east and west; the Wenyu River flows to the south. It is, therefore, a perfect example of a propitious Feng Shui place. In this vast landscape, the main built elements were easily accommodated. The first that one encounters, as in Xiaoling, is the magnificent Sacred Way, which also serves as the visitor's entrance to all the subsequent tombs. The path is very long (more than 7 km in total) and starts with an enormous (12 m high, 29 m wide) 5-gate stone archway (added probably in the sixteenth century) followed by a large, roofed gate (Great Red Gate). The Red Gate was a real entrance in the sense that it gave access to the perimeter wall encircling the whole area, of which almost nothing remains. In front of the gate, two paired stelae invite all visitors to dismount from their horses. Further north, a pavilion houses the monolithic tortoise stela (height about 8 m) of "divine merits", which bears an inscription dedicated to Yongle. The tradition of adding this impressive monument, which started at Xiaoling, was to be followed for all the tombs of the valley. Finally, the statuary of the Sacred Road starts, extending for about 1 km. First, the monolithic pillars (engraved with clouds) then the animals, both real and mythological, all in pairs, and all shown once standing and once crouching. Finally,

Fig. 7.1 The Thirteen Ming tombs, numbered in chronological order. (1) Yongle (Changling) (2) Hongxi (Xianling) (3) Xuande (Jingling) (4) Zhengtong (Yuling) (5) Chenghua (Maoling) (6) Hongzhi (Tailing) (7) Zhengde (Kangling) (8) Jiajing (Yongling) (9) Longqing (Zhaoling) (10) Wanli (Dingling) (11) Taichang (Qingling) (12) Tianqi (Deling) (13) Chongzhen (Siling). *Credits* Images courtesy Google Earth, editing by the author

military court officials with helmets and swords, civil officials (with simple hats not covering the ears) and noble officials with elaborate hats (Figs. 7.2, 7.3, 7.4, 7.5 and 7.6).

Interestingly, as in Xiaoling, after a first straight section, the path bends, to confuse malevolent spirits who "go straight". At the end of the path, the wonderful "Dragon and Phoenix Gate", painted red with glazed tiles and the carving of what seem to be flaming pearls (whence the alternate name, "Flame Gate") but may also represent peaches (an auspicious fruit for Taoists) brings the visitor to a sequence of two bridges, at the end of which, the tomb proper begins. Intriguingly, the monumental axis of Changling proper is not parallel to the main section of the Spirit Road, but points to a very small hill which was, in fact, the only available hill on the southern horizon.

The tomb is accessed by a front gate followed by a further stela pavilion, probably added during the Qing dynasty. The main part of the mausoleum is then entered through the *Gate of Eminent Favor*. After this gate, we see at the sides two strange miniature temples decorated with glazed green and yellow tiles. They are actually ritual owens where pieces of paper calling for good omens were burned. In front of the visitor, at last, appears the greatest masterpiece of the Ming architecture in the Valley: the *Ling-en Men* or *Hall of Eminent Favor*. It is essentially an offering hall, where memorial tablets were held and sacrifices were made in memory of the

Fig. 7.2 Sacred Way of the Thirteen Ming tombs. Stone camel

Fig. 7.3 Sacred Way of the Thirteen Ming tombs. Stone lion

Fig. 7.4 Sacred Way of the Thirteen Ming tombs. *Quilin*

deceased. Built on a three-tier platform of white marble, it is a vast open space, accessed by three stairways, the central one of which was clearly reserved for the emperor, engraved as it is with a fine relief of Dragons and Mountains. The whole complex strongly recalls the Hall of Supreme Harmony in the Forbidden City: the structure rests on sixty single-trunk columns of nanmu hardwood, more than 12 m high (inside, a large but modern statue of the emperor helps us to remember that we are not traveling in time and space) (Fig. 7.7).

After the Hall, we enter the final part of the complex through a triple-opening gate. In the courtyard, we further pass another symbolic (no walls on the sides) door, the *Lingxing Gate*, resting on two pillars with auspicious animals on their top (it should be remembered that evil spirits hate gates). After this gate a raised platform or altar bears five sculptures of sacrificial vessels (one incense burner, two flower vases and two candlesticks). In front of us, finally, appears the Minglou—the Soul Tower, the front wall of the mound—topped by a stela pavilion covered with a double-eaved roof with yellow glazed tiles. The enormous stela inside bears the name of the emperor. From the pavilion, looking north, we can see the circular mound under which the owner was entombed along with Empress Xu.

All in all, Changling was conceived on an almost unprecedented scale, which can only be compared to the most sumptuous of the Tang projects, Qianling. However, while in Tang tombs the open perspectives facilitate the intertwining of the two worlds, in Ming tombs—as had already occurred in Xiaoling, but was established definitively here—the part reserved for the living, with buildings and rectangular courtyards, is clearly separated from the circular mound and—ideally—from the underground palace. The transition is announced by the Soul Tower: actually a huge

Fig. 7.5 Sacred Way of the Thirteen Ming tombs. Military officer

Fig. 7.6 Sacred Way of the Thirteen Ming tombs. Civil officer

Fig. 7.7 The Hall of Eminent Favor, Changling

wall so similar to the walls of a strongly fortified city (such as those of Ming Nanjing), but with gently-sloping stairways ascending to the top. Ceremonies and celebrations in honor of the dead were held daily in the "living" part; since, of course, the ruling Emperor took part in the most important of them (on the days of the dead festival and in other yearly occurrences) the architecture of this part had to be on a par with imperial palatial architecture to the point that, as mentioned, the Sacrificial Hall at Changling is a almost a replica of the Hall of Supreme Harmony in the Forbidden City.

7.2 Thirteen Emperors, Thirteen Tombs

The use of a royal necropolis previously established by predecessors has profound political and symbolical significance: indeed, as we have seen, sacred places tend to become permanent in cultural memory. Examples can be cited from very different civilizations and epochs, from the Valley of the Kings at Luxor, Egypt, to the Abbey of St. Denis in Paris, France. The Valley of the Thirteen Tombs is no exception. Imperial unity and coherence of intention of the ruling family were rendered explicit here by the addition, one after the other, of the tombs of *all* the Ming emperors. The projects of the tombs were planned on the basis of the existing Sacred Road, and today they radiate from it like the branches of a tree (of these tombs, only one—Dingling—has been excavated; we shall visit its interior in next section) (Fig. 7.8).

Fig. 7.8 The Valley of the Thirteen Ming tombs in a seventeenth century painting. *Credits* Images in the public domain

As a general observation, it should be noted that each single project had to satisfy as far as possible the overall inspiring principles of landscape planning adopted at the court, that is, Feng Shui. The architects accordingly had to face the problem that the best of all positions was already taken by Changling, which is nested at the very end of the valley and at the very foot of the main mountain. However, an important characteristic of the site is that practically every subsequent tomb could be built in such a way as to have a "principal mountain" in its background: this requirement clearly governed orientation, and, indeed, there is no evidence of the use of compass (or astronomical) orientation. This is confirmed quantitatively by the linear correlation test between orientations and palaomagnetic declinations (Table A.5), which returns complete absence of correlation ($R = 0.05$). Visiting the valley, however, it is evident that some royal architects were more successful than others in placing the tombs of their rulers. We shall now describe the evolution of the site over time, and then concentrate on one of its masterpieces, Dingling.

The second tomb of the valley is that of Yongle's son Hongxi. The emperor reigned for only one year and his tomb was planned to be fairly modest; he probably expressed the wish to be as close as possible to his father, but the place chosen is still particularly fascinating. It is a sort of small valley immediately to the northwest of Changling, which is also enclosed—unique among the tombs of the valley—from the south by a low hill. The tomb was then designed in a perfect Feng Shui position with respect to both these "mountains". There was insufficient room to insert all the elements in the tomb valley, though, so the gate and the sacrificial area were constructed further to the south. The main axis of the tomb points to the stela Pavilion of the Sacred Road.

The third tomb, that of Emperor Xuande, was built to the east of Changling (the funerary mound of this tomb is oval in shape rather than round). The first three tombs actually form a symmetrical, quite intriguing, ensemble. For the subsequent three tombs, the decision was taken to move progressively to the north-west (the first free area to the northwest is however occupied by Tomb 11, for a reason explained below). For Tomb 7 (Kangling of Emperor Zhengde), the architects had to move away from the sides of Mount Zhunua and to build the tomb at the foot of Mount Jinling, on the west border of the valley. As a result of the topographical situation, the tomb could not be oriented to the southern horizon. The result is, nonetheless, spectacular: the axis of the tomb crosses the tortoise stela and proceeds undisturbed through the whole valley, passing over Changling and ending at a small peak some 4 km to the west.

The Emperor Jiajing project followed. As we saw in the previous section, he had a rather curious personality and was particularly preoccupied with his Taoist beliefs. It appears that he personally made many choices for his tomb, starting from the location, since it is located almost in the plain to the south of Changling. The axis was carefully chosen so that Mt. Tianshou is in the background as a "Crouching Dragon" and a small hill is targeted to the south west. As a consequence of its owner's beliefs, the tomb has many peculiarities; for instance, it was walled separately, with its own triple gate access, and the interior sections were walled too (another curious detail is the presence, on the square stone platform of the tortoise stela, of carvings of marine creatures: a shrimp, a crab, a turtle and a fish). In spite of its magnificence, however, this is by no means the tomb to which this Emperor directed his most efforts. As mentioned in the previous Chapter, he was obsessed by the legitimization of his power directly from his parents, and so he took great pains to bring Xianling, the mausoleum of his father, to perfection, with the aim of transforming it from a kingly mausoleum (the status he had at his death in 1520) to an imperial one. The result is by far the largest Ming tomb ever constructed, and it is located in the Songlin Hills, northwest of Zhongxiang. The tomb escaped most of the destruction to which the Ming Emperors' tombs have been subjected: it still retains, for instance, its perimeter wall, which is not straight, but curves following the steep of the hill. Another distinctive characteristic is the masterly use of the water of river Yuhe. The river meanders through the entire area, criss-crossing the Sacred Way several times, and is spanned each time by a bridge. Water also forms two ponds, a "one-off" among the Ming tombs. The first, and larger, has a cardiod shape and is located near the entrance; the second is circular and located in front of the Soul Tower. Curiously, the axis of the tomb ends with two burial mounds in sequence, connected by a terrace. The second one was constructed in 1538, when Jiajing Emperor's mother died.

With Longqing Emperor—Tomb 9—and Wanli—Tomb 10—the Ming funerary monuments return to the west side of the valley. The tomb of Longqing enjoys a position at the western end and points to the same heap of the axis of Changling. Dingling, the tomb of Wanli, will be discussed in next section.

After Wanli, the Ming Dynasty was doomed to a rapid demise. His son and successor Taichang died after only one month of reign. His tomb—number 11—re-utilized the underground structures which were close to being finished for the tomb of Jingtai Emperor, the "would-be" fourth emperor to be buried in the valley, who was degraded to noble status and buried in a Prince's tomb (the tomb of Jingtai is in good condition and is located on the western hills of Beijing suburbs). This is the reason why this tomb is close to the first three, and in particular is in an alternate position with respect to Tomb 3. Wanli's grandson Tianqui reigned for six years, and his tomb (n.12) was located in a rather awkward position in the immediate vicinity of Jiajing's. The last Ming Emperor Chongzhen committed suicide in 1644. The new rulers, the Qing, ordered his body to be interred in the valley, and it is reported in some sources that the Emperor was buried in an pre-existing tomb, that of his concubine Tian. However, the topographical siting of this tomb is curious, as it is different not only from the imperial tombs but also from the other tombs of the Ming concubines. The latter—today almost ruined and difficult to find among the cultivated fields—are located at the foot of the hills to the southwest of the imperial area and oriented topographically with respect to these hills, while Chongzhen's is located on flat plain and is quite precisely oriented to geographical north (a choice which seems appropriate for an emperor). So perhaps this tomb was constructed anew, with a project which was as a compromise between an official emperor's tomb and a secondary tomb, locating it far from the other imperial tombs, but orienting it to true north.

7.3   A Palace for Eternity

One of the projects of most impressive beauty within the Ming valley is Dingling, the tomb of Wanli (1572–1620). This tomb has been carefully restored and, moreover, is the only one that has ever been opened. Wanli was the thirteenth Ming Emperor. During his long reign, the dynasty verged on collapse, mostly because of his own lack of interest in state affairs. Indeed, the inefficacy of the army rendered the borders unsafe, both in the north (with the Mongols) and in the northeast. Here, in particular, the Manchu chieftain Nurhachi started to reorganize the Manchurian tribes into a formidable war machine, rapidly leading—as we shall see—to disastrous consequences. Meanwhile, the emperor was more involved in planning his own tomb, which, according to historical sources, took six years and required a tremendous amount of resources. The result, however, is distinctive for the particularly ingenious combination of natural and built elements its builders were able to achieve (Fig. 7.9).

The Dingling layout is not fundamentally different from Changling, and is arranged on a straight axis with a succession of three courtyards. The first element is a bridge crossing a small river. Then the tortoise stela (probably originally housed in a square building, now destroyed) decorated at the top with playing dragons. Then a large platform leads to the three-opening front gate, with the spectacular mass of Mount Dayu in the distance. The buildings of the interior courtyards were mostly destroyed at the end of the Dynasty and only their stone foundations remain. The Back Courtyard has, however, been restored with the magnificent double-pillar "spirit" gateway, or *Lingxingmen,* and the five sacrificial stone vessels; at the end, the Soul Tower is accessed via two symmetrical ramps and contains a huge stela.

Fig. 7.9 Dingling, front view

An impressively beautiful straight line can be seen connecting, from west to east, the summit of the protective hill, the axis of the tomb and the tortoise stela, and then extends up to the summit of the opposite hill, which is marked by a pagoda. The axis does correspond to solar azimuths at rising (with dates around end of January/mid November) but this is probably purely coincidental, as the intention of a topographical alignment is undeniable. Incidentally, since this is the only imperial tomb to have been excavated, we know for certain that the axis of the corridor of the underground palace also runs in the very same direction (Figs. 7.10 and 7.11).

The history of the excavations at Dingling—carried out in 1956—is not crystal-clear; a film however is available which shows various stages of the work. The funerary chambers lie exactly where one would expect, namely, in axis with the tomb, under the mound. So anyone wanting to access the tomb would have dug along the axis under the mound starting from the front of it. However, apparently, the excavators found themselves in difficulty, until they found a tunnel under the Soul Tower, and at the end of that a small stone tablet came to light. The text on it gave information about the distance and the difference of level "from this stone to the front

Fig. 7.10   Dingling, looking from the front towards the tortoise stela

Fig. 7.11   The Feng Shui alignment at Dingling: (1) Summit of Mount Dayu (2) Tomb mound (3) Soul Tower (4) Tortoise Stela. *Credits* Images courtesy Google Earth, editing by the author

of the Diamond Wall" as (in modern measurements) 52 × 11.5 m. What "Diamond Wall" was supposed to mean by the person who wrote the text will soon be clear, but the reason why the builders left this "directional tablet" after the definitive sealing of the tomb remains a true mystery (at least to me). In any case, the information proved to be quite correct. In fact, at the predicted distance, the excavators found a curious wall of stone bricks. The images of the discovery show a continuous wall where it is pretty clear, though, that a A-shaped opening had been refilled by bricks. Once this fill was removed, the A-like entrance—which might easily be defined as a "Diamond Wall"—revealed an empty antechamber, housing the immense marble doors of the entrance to the tomb (Fig. 7.12).

Emotion must have been palpable in the air, since it was quite clear that the tomb was intact. The doors were perfectly sealed and without any handles: they had been blocked from the inside by means of a sliding mechanism which caused a marble pole to slide under a small protruding bar on the backside of the door, making it impossible to push open the door again from the outside. Once opened the doors— by removing the pole through the insertion of a wire—the excavators finally entered the untouched underground palace—as we are now going to do as well.

Entering the funerary apartments of Dingling is an awe-inspiring experience. They were built inside a huge pit, up to 26 m deep. The pit was paved and encased in stone, with interior walls and corridors being built to create what was virtually an enormous box with internal compartments. Then the megalithic marble doors were laboriously

Fig. 7.12 Dingling. The diamond wall

lowered and fitted onto their hinges, and finally the whole underground palace was covered by a vaulted ceiling. The pit was then filled in with earth and covered by the mound, leaving only the entrance tunnel—also to be filled in after the funeral. The apartments take up more than 1000 m^2 and have a strict, formal structure, which in a way recalls the rigorous spatial hierarchy of the Forbidden City. Indeed, behind the marble door and in axis with it, there is a large corridor, 6 m wide and 32 m deep; parallel to this run two secondary corridors—or rectangular chambers—connected by galleries to the central one. Therefore, the underground palace is tripartite in the longitudinal direction, with the main section reserved for the emperor. The lateral chambers were found empty, while the central hall houses three marble thrones, like imperial spirit seats, each equipped with ritual vessels and a blue porcelain vase, probably used as an oil lamp (Poo 2011) (Fig. 7.13).

The burial chamber—analogous to the private part at the end of the Forbidden City—is at the end of the main hall. Inside, over a stone platform, the coffins (today replaced by copies) of Wanli and of Empresses Xiaoduan and Xiaojing were placed, together with lacquered chests of nanmu wood filled with the burial equipment. During the excavations, the opening of the coffins, and the inspection of the skeletons and of the boxes was evidently done in something of a rush, as shown in the striking images available (all in all, the tomb yielded more than 3000 objects, including the golden crowns of the tomb occupants).

Fig. 7.13 Dingling. The main hall in the underground palace

To conclude, we can say that the use of Form Feng Shui is apparent in all aspects of the thirteen Ming Tombs complex, starting with the choice of the site and continuing throughout the entire evolution of this sacred landscape to the point that, with a little imagination, we can visualize the imperial geomancers and architects studying, case by case, the optimal locations for the subsequent tombs in the valley. Given such a magnificent result, at the end of the Ming Dynasty a tricky issue had to be addressed by the subsequent dynasty: how to find an equally auspicious place for their own Necropolis.

As we shall now see, the dilemma was brilliantly solved.

Chapter 8
The Last Dynasty

8.1 A New Geography for China

The fall of the Ming Dynasty was heralded by fierce peasant revolts, which gave rise to the conquest of Beijing and the suicide of Emperor Chongzhen. The road was thus paved for a second period of foreign domination after the Mongols: the Qing Dynasty of the Manchus, originating in the northeast, beyond the Great Wall. Under the tribal chieftain Nurhachi, they had managed to create a unified state on their own; in 1616, Nurhachi conquered Shenyang and elected it as his capital; from here, he embarked on the subjugation of the whole country. The original Ming town of Shenyang was re-arranged to accommodate the structure of the Manchu tribes and army, which were divided into eight "banners", with the residence of the chief in the center. The Manchu capital rapidly grew in importance and, in accordance with the faith in Tibetan Buddhism of Nurhachi's successor, Hong Taijii, four pagodas were constructed along the four cardinal directions in relation to the royal palace, in a manner similar to the four temples built by the Ming in Beijing.

The establishment of the Qing rule was a rather lengthy process, concluding under Shunzi (1644–1661), who was actually the first Qing Emperor to rule over the whole of China (Wakeman 1985). Shunzi was proclaimed Emperor on the fall of Beijing—with a series of official ceremonies including sacrifices at the Temple of Heaven—when he was only six years old. Prince Dorgon (one of Nurhachi's sons) was appointed regent and effectively ruled China for seven years, during which the war of conquest reached completion and some decisions (including some extremely unpopular ones) started to be taken to favor the integration of the two populations. In particular, the Chinese were forced to shave their foreheads and to wear pigtails in the Manchu style, a rule which occasioned much protest, especially among strictly observant Confucians. When Shunzi became the effective ruler at age thirteen, however, he gave a decisive boost to the integration of the country precisely by building on traditional Confucian principles. In this way he imposed an order that, although dominated by the Manchu ethnic minority, was acceptable to the Chinese people as a whole.

G. Magli, *Sacred Landscapes of Imperial China*, https://doi.org/10.1007/978-3-030-49324-0_8

His reign was not destined to be a long one, however, as the emperor contracted smallpox (against which, apparently, Manchu people had no genetic immunity) and died at the age of twenty-two. Shunzi was succeeded by one of the most brilliant and long-reigning emperors in Chinese history, his son Kangxi (1661–1722). Under Kangxi, there began a period usually called the "Prosperous Era" which lasted up to the end of Qianlong reign. Besides military conquests and economic success, Kangxi is renowned for the enormous impact he had on culture. In particular, he ordered the compilation of collections of Chinese poems and of a complete dictionary of Chinese characters. The main additions to Beijing carried out by Kangxi and his immediate successors were the two imperial residences, traditionally called the Old Summer Palace and the Summer Palace, both located in the northwest suburbs of the capital. The Old Summer Palace or *Yuanmingyuan*—Garden of Perfection and Light—was designed in 1709. It was made up of three gardens, boasting "scenic spots" created with charming pavilions, halls, kiosks, streams and ponds. Visiting it today is a rather bizarre experience because in 1860, during the Second Opium War, the Anglo-French army commanded by Lord Elgin—in revenge for the treatment accorded to their delegates—looted and virtually destroyed the place. In any case, the true masterpiece of Qing architecture in Beijing is the Summer Palace. The project, developed in a vast area where temples and palaces already stood, was based on an existing lake, which was substantially expanded by digging out two other large, artificial ponds. The resulting artificial reservoir, known as Kunming Lake, contains three artificial islands, meant to represent the three mountain islands of the east traditionally associated with immortal lands (the same mountains that Shihuang had been seeking some 2000 years before) (Luo and Grydeho'j 2017). The excavation of the lake produced a massive amount of earth, which was used to consolidate and enlarge a hill located on the northern shore. The hill took the name of Longevity Hill and has a number of buildings on it, many of them inspired by existing ones. So again we have the reappearance of the idea of building a landscape as a microcosm representing mythical places or real parts of the Empire, to say nothing of the notorious, extravagant additions made by Empress Dowager Cixi in 1893, which included a western-style marble steamboat. In any case, the most complete and complex of such microcosms is by far the one conceived by Kangxi in a place called Chengde (Berger 2003; Foret 2002) (Fig. 8.1).

To understand Chengde we must keep in mind that Kangxi devoted considerable effort to constructing and consolidating a Qing ideology of power based on the idea of a united Chinese/non-Chinese people, bringing together Han, Manchu, Mongols (many of whose territories became annexed to the empire) and Tibetans. The spectacular building program of Chengde was contributory to this effort. Usually called a "mountain resort" (but this definition will be shortly updated), Chengde (also called Jehol) is a hilly, green countryside, some 250 km to the northeast of Beijing, outside the Great Wall. In 1703, the Emperor chose this place for the building of a majestic imperial residence and launched an impressive series of constructions, which reached its apogee in the reign of Qianlong, around 1780. The vast complex is organized thus: the main entrance to the imperial palaces is due south, and the palace structures develop along a strict south-to-north axis. To the east, an intricate

Fig. 8.1 Summer Palace, Beijing. View from the north-east along Kunming Lake, with Longevity Hill on the right

series of small artificial lakes—connected by artificial islands housing pavilions and palaces—creates a peaceful landscape. To the north, immersed in the vegetation, a series of other buildings and pavilions were built for conducting state affairs and hosting important guests. Seen from the air, the huge, verdant area of the palace can be seen to be bordered by a series of buildings to the east and the north. These are the so-called outlying temples, twelve sacred places (mostly constructed under Qianlong) inspired by, or even conceived of, as replicas of significant Buddhist temples. Among these, two stand out and take visitors' breath away: the Putuo Zongcheng Temple and the Pule Temple (Figs. 8.2 and 8.3).

The Putuo Zongcheng Temple, built in 1767 and located due north of the imperial palace complex, was modeled after the architecture of the Potala Palace, the winter palace of the Dalai Lamas built in 1649 in Lhasa. Comprising dozens of buildings scattered about a natural hill, it culminates in an enormous red-and-white palace built on the summit, and the fusion of architecture and natural features really does give the impression of being in the Tibet. Another particularly striking building in Chengde is the Pule Temple, again built by Qianlong. The temple was built to the east of the palace area, in a panoramic position that gives a view, along an axis almost due east, of a very special natural feature, the so-called Qingchui peak (seen from the temple, the sun rises in alignment with the peak a few days after the Spring equinox and a few days before the Autumn equinox). The peak is a large, oddly-shaped rock formation, resembling an upside-down hammer, and it has been suggested that the presence of such a peculiar peak may have inspired the choice of the entire area (Foret 2001). Whatever the case, the alignment of the temple with the peak is evident, and the architecture of the temple itself is quite curious: it is built on three huge square terraces, and the main building clearly resembles the Hall of Prayer for Good Harvests of the Temple of Heaven in Beijing. A final notice deserves the Puning Temple, to the north-east of the palace area. Built in 1755, it houses a wooden

Fig. 8.2 Putuo Zongcheng Temple, Chengde

Fig. 8.3 The Pule Temple with Qingchui peak on the horizon, Chengde. *Credits* Images used under license from Shutterstock.com

statue of Avalokitesvara, a 1000-armed Bodhisattva (a personification of Buddha), embodying the compassion of all Buddhas. Usually this statue is defined as "large", and this is no exaggeration—it stands at 22 m high.

It should thus be said that the project devised by Kanji and brought to completion by Qianlong in Chengde is something quite different from a "mountain resort". The Qing Emperors did, in fact, transfer to the temperate climate of Chengde from sultry Beijing regularly in the summer, and there they held court, received state visits and went hunting. However, the choice of the place and its architectural development was imbued with a profound and interlinked political, symbolical and religious significance: Chengde is a sacred space conceived of as a microcosm of the Qing empire, the powerful dynasty that united China and Central Asia under the Manchu rule. In a similar way to the first emperor, who had copies of the palaces of his defeated enemies built in his capital, building replicas of major architectural projects was a way of affirming the completeness of the Qing unification. Politically, the "mountain resort" reproduced the empire, while, from the religious point of view, the geography of the site explicitly referred to Tibetan Buddhism. In addition, the "mountain resort" played a pivotal role in what may perhaps be considered the most astounding application of Feng Shui ideas to political propaganda in the whole history of China (Whiteman 2013; Harley and Woodward 1995).

In 1718, Kangxi requested a complete survey map of the country, today known as the Kangxi Atlas. The aim of this huge cartographic endeavor was eminently political, as it helped to open the mind and the attitudes of the Chinese to the extended geography of the empire, dissolving geographic distinctions between the recently fallen Ming borders and the northeastern and Central Asian territories enclosed within the Qing multi-ethnic country. Interestingly, this atlas also included ritual and ideological components. In point of fact, Kangxi conceived and created a "geomantic meta-geography" that mirrored the established Chinese conceptions—in particular, that of the sacred land bordered by the four-plus-center sacred mountains—extending these conceptions to the new borders. This was effected through texts but also through explicit acts. In particular, the Emperor relocated the ancestral mountain ritual of the Fengshan from Mount Tai to the Changbai Mountains and the peak of Changbaishan, which was the mythical homeland of the Manchus. In doing so, he elevated this mountain to the same status as the Five Sacred Mountains of Chinese tradition. In this relocation, the "mountain resort" happened to be in the correct place to convey "geomantic currents" from the north, acting as a pivot for the new geography. Later, he went even further than this. In the summer of 1677, Kangxi demanded a complete mapping of the Changbaishan mountains, and—not surprisingly—the report he received emphasized the auspiciousness of the peak and of the surrounding landscape. The Emperor thus commanded rites to be conducted there with the same protocols as those of the other five sacred mountains. The propitiousness of the site was further confirmed by the Emperor in person: Kangxi wrote an essay whose title could not be clearer: "Taishan Mountain Veins Originate in Changbaishan". In the essay, which is a masterpiece of Feng Shui adaptive reasoning, he claims that the source of the Empire's geomantic force is Changbaishan. Essentially, he applies the standard analysis of mountain veins (dragon to the east, tiger to the west and so on) to the whole of China, identifying the Dragon protective hill with Taishan. The importance of Taishan is thus stressed, but the question is: where do its Dragon veins really come from? And the answer is Changbaishan. As in any respectable Feng Shui analysis, the idea is elaborated with a closer look at mountain ranges and watercourses, and corroborated by the presence of the tombs of the royal ancestors to the north. In short, Taishan is the first among the five sacred mountains of China, but Changbaishan—sacred to the Manchu and therefore identified with the reigning Manchu dynasty—is its true source of its vital energy.

8.2   From Shisanling to Fengtaling

The Qing efforts to legitimize their Mandate of Heaven naturally included the adoption of ancestor worship and the complex ceremonies associated with death and burial. A reflection of this can be seen in Qing architecture and, in particular, in the two spectacular necropolises they built. However, the first architectural projects were devoted (as soon as the domination over China had been stabilized) to the abovementioned tombs of the first two Manchu Emperors and of their paternal ancestors, to whom posthumous imperial status had been conferred. This led to the construction of three imperial tombs north of the Great Wall. The earliest of these tombs, Yongling in Fushun, clearly exhibits the Manchu's traditions and style. It was built for the ancestors of Nurhachi (1559–1626) (who also built a further tomb, Dongjingling in Liaoyang, for some other of his relatives). In the subsequent tombs—Fuling, the tomb of Nurhachi, and Zhaoling, the tomb of the second Qing Emperor Hong Taijii, both in Shenyang suburbs—additions conforming with Chinese traditions, such as sacred roads, were made, although the tomb complex retained the appearance of a feudal castle. It is difficult to ascertain whether the Qing wanted to follow the Feng Shui principles already in these early tombs; certainly, in all three cases a river flows to the south and the axis of the complex is roughly perpendicular to the river (Table A.6).

Once the dynasty was firmly installed, the Qing rulers had to select a site for their own Necropolis At the same time, they wanted to give a strong signal of continuity with regard to the Mandate of Heaven and Chinese burial tradition, and so—apparently—Emperor Shunzi himself studied the principles of Feng Shui. Actually, according to some historical records it had already been the last Ming Emperor Chongzhen who had tasked the imperial geomancers with finding a favorable new place for his tomb, and they suggested the site of Fengtaling, some 120 km northeast of Beijing. Be that as it may, Chongzhen died before he could start building there. When Shunzi visited Fengtaling, he confirmed the auspiciousness of the site and inaugurated the Necropolis today called the Qing Eastern Tombs, where in total four Emperors, four Empresses, and 150 other members of the royal family are buried (Wang and Li 2014; Brook 1989) (Fig. 8.4).

There is absolutely no doubt that Feng Shui was the inspiring principle behind the choice of Fengtaling; however, the place is very different from the valley containing

Fig. 8.4 The imperial tombs and their satellites in the Qing Eastern tombs area, numbered in chronological order. (1) Shunzhi (Xiaoling) (2) Kangxi (Jingling) (3) Qianlong (Yuling) (4) Xianfeng (Dingling) (5) Tongzhi (Huiling) (6) Unfinished (Daoguang?). North is slightly skewed to the right to enhance the topographical features. *Credits* Images courtesy Google Earth, editing by the author

the Ming tombs. Fengtaling is, in fact, a vast flat area of land, located roughly to the south of a straight-front line of hills, the Yanshan chain. The chain is shaped in a "double dragon" profile, with the main peak (Mount Changrui) approximately at the center. To the south, a peculiar-looking "protective" hill, Mount Jinxing, can be seen. It stands alone on an otherwise flat horizon, in rough north-south alignment with Mount Changrui, and bears—perhaps by chance—a certain resemblance with the peak of the mausoleum of Taizong, the first Tang Emperor who choose a mountain for his tomb. It is easy to imagine that, when the imperial geomancers first saw this place, they immediately visualized a spectacular axis traversing the whole site, with Shanzi's tomb (called Xiaoling) positioned right at the foot of the main peak, on the focus of the axis. The result is a perfect Feng Shui alignment running along a gigantic tomb complex. The Spirit path is not entirely straight, in order to meet "magical" requirements, but a line does run straight along the axis of the tomb, the stela pavilion, and a relevant section of the sacred road (Fig. 8.5).

Overall, the whole complex is very similar in conception to Changling, the first project built in the Ming Valley of the Thirteen Tombs. Entry is via a magnificent, five-gated stone arch where visitors are, of course, invited to dismount from horses. Crossing the arch and looking south, the "mountain of the south" is framed scenically. The Spirit Path, flanked by eighteen pairs of stone animals and statues of officials, is also very similar to the Ming one, the only difference being that the statues of officials are dressed in Manchu style and wear the typical pigtail that the Manchu compelled the Chinese to sport as proof of their loyalty (Figs. 8.6, 8.7 and 8.8).

At the end of the road, a huge arch introduces the next section, with the "Dragon and Phoenix Gate" and the Pavilion of the Tortoise Stela. Further north, there are ceremonial bridges and the entrance to the Long'en Palace, conceived as usual as

Fig. 8.5 The main axis of the Qing Eastern tombs (north to the left): (1) Xiaoling (2) Main bridges (3) Ceremonial gates (4) Bend in the Sacred Way (5) Stela Pavilion (6) Mount Jozhong. *Credits* Images courtesy Google Earth, editing by the author

Fig. 8.6 The Sacred Way of the Qing Eastern tombs, looking south with Mount Jozhong at the horizon

the transitional place of encounter between the living and the dead, where memorial services were held. Finally, a five-piece stone altar marks the approach to the Soul Tower, in front of the burial mound, encircled by a spectacular, curved wall. Inside the tower, a stone stela bears the name of the tomb (Fig. 8.9).

Contrary to what happened with the Ming, who reopened the tombs in the case of Empresses surviving their husbands and buried the concubines in a specially chosen, separate part of the Necropolis, the Qing Empresses were buried in independent mausoleums located not far from the main one, and the concubines were buried in cemeteries also not far from the main one. Each imperial Qing tomb thus had one

Fig. 8.7 Stela pavilion, Xiaoling, Qing Eastern Tombs

Fig. 8.8 "Tortoise" stela with dragon head, Xiaoling, Qing Eastern Tombs

Fig. 8.9   Altar of the five stone objects and Soul Tower, Xiaoling, Qing Eastern Tombs

or more satellites, so that the Qing Necropolises generated an impressive number of monuments (Table A.7).

After Shunzi, Kangxi was also buried in Fengtaling. After the massive, spectacular Changling project had been implemented, however, the place was no longer specifically suited for the proper geomantic placement of other tombs. Nevertheless, Kangxi successfully placed his burial complex, Jingling, very near to Xiaoling, to the east. The project of this tomb, though very beautiful and monumental in its own right, clearly demonstrates Kangxi's wish to express his close connection with his father, as the road for the tomb radiates after the second monumental bridge of the Xiaoling access way, and the two tombs are located at less than 1 km as the crow flies.

The succession of Kangxi was problematic. The dissolute behavior of the Emperor's heir apparent, Yinreng, made him unworthy of the succession, and Kangxi placed the name of the new heir he designated inside a box to be opened after his death. The name was that of the fourth prince, who ascended to the throne as Yongzheng. His reign was inspired by rigorous Confucian doctrines; in spite of this, many rumors circulated about his forging the text of the succession papers to the detriment of one of his brothers. According to some, this forgery was the reason why he feared to be buried near his father. However, these rumors have been dismissed by historical researchers, and, in fact, we do not know why he decided to break the tradition established by the two previous emperors by building his tomb in a completely new place, founding the Necropolis today known as the Qing Western tombs. In the existing Necropolis of Fengtaling, however, he did order the construction of one beautiful tomb. It is located down in the valley and beyond the large red gate, with the axis almost aligned with the southern mountain, which in that area is so close as to dominate the landscape completely. The tomb belongs to Empress Xiaozhuangwen, Shunzi's mother, who died in the 26th year of the reign of Kangxi (1687), after a period in which she had exercised a strong influence on the state affairs. She expressed the wish to be buried near her son. The decision was finally taken by Yongzheng to respect her will by constructing her tomb in Fengtaling, but in a displaced position; the tomb was called the Western Zhaoling Mausoleum (Fig. 8.10).

In the site today known as Qing Western Necropolis, Yonghzen ordered the construction of a tomb which is in many respects similar to Chongling. Thereafter, both places—the Eastern and the Western—continued to be used, so that there are two Qing Royal Necropolises, and it is essential to follow their development together. However, describing the two places alternately would not be helpful in understanding the evolution of the single landscapes, so I shall first describe the progression of the building projects of the Eastern Necropolis, mentioning the projects of the alternate site in chronological order; thereafter we will turn our attention to the Western Necropolis.

The successor of Yongzheng was his fourth son Qianlong. He built his tomb in the eastern necropolis, but—according to historical sources—from then on he wished burials to alternate between the two sites. His tomb, Yuling, is thus the third imperial tomb built in Fengtaling. It is located to the west of Xiaoling, a natural choice, since

Fig. 8.10 Western Zhaoling, Qing Eastern Tombs. *Credits* Images used under license from Shutterstock.com

the first position to the east was occupied by the second project, that of Kangxi. Also in this case, the road for the tomb radiates from the Xiaoling sacred road as a branch, and the main axis of the tomb complex is oriented towards the peak of the southern mountain. To the immediate west of the complex lies the mausoleum of Qianlong's concubines, a fairly impressive sight with a main burial site (with Soul Tower) and a sort of orderly forest of identical tombs covered by cylindrical brick mounds in the backyard.

Qianlong's is the only one of the emperors' tombs of the Eastern Necropolis to have been opened, and this occurred in dramatic circumstances. Indeed, on June 12, 1928, the army of the warlord Sun Dianying took possession of the area and looted some of the tombs. The tomb of Emperor Qianlong was the first to be blasted open and violated; later, other tombs, including the tomb of Cixi (see below), were also opened. Nothing was spared, including the objects lying with the corpses of the deceased. The underground palace of the tomb of Qianlong, today restored and accessible, consists of three rooms in a row, whose interior marbles are finely carved with Buddhist reliefs. A little known feature is the presence of inscriptions relating to Tibetan Buddhism. Qianlong's espousal of this religion as ruler is, of course, beyond doubt (one only has to think to the temples he added to Chengde), but he is also known for his public adoption of Confucian traditions. The inscriptions in the intimate solitude of his own tomb clearly attest to a full, personal embracing of Buddhism. These inscriptions, engraved on the walls and the coffin, are written in Tibetan and in Lantsa, the Tibetan Sanskrit (they are currently at risk due to the high humidity in the tomb, a result of modern hydraulic works in the area). The texts engraved on the walls and the vaults are only sacred formulas, while those on the coffins also contain prayers. A recent analysis of these texts (which were mostly composed for the sound they produced when read, rather than for their actual content) has revealed a surprisingly complex organization of the sacred space in the rooms of the tomb. The texts were carefully selected and divided in accordance with the different rooms. In the first chamber, protective divinities are engraved on the walls, together with texts that are well-known for their apotropaic function. Texts linked with consecration rituals are engraved in strategic places around the rooms, such as vaults and lintels. Texts of purification and rebirth, associated with Buddhist funerary rituals, follow. All in all, the distribution of these texts is a sort of "photograph" of a Buddhist funerary ritual. In addition, the final chamber has a surprise in store. The deposition of important relics, and of texts in particular, was a very important practice in Tibetan Buddhism. In particular, specific texts were associated with each of the architectural parts of the *stupa*, the Buddhist funerary monument, in which they were placed. The inscriptions in the burial chamber of Qianlong appear to be related to the same distribution of texts, thereby forming a sort of virtual *stupa* containing the coffin of the emperor as a relic (Wang and De Luca 2018) (Fig. 8.11).

After a very long reign, Qianlong abdicated in favor of his son, Jiaqing, whose choice was to respect his father's will. His tomb is accordingly located in the Western Cemetery. His successor, Daoguang, was thus expected to build his tomb in the

Fig. 8.11 Antechamber of the underground Palace, with finely carved Buddhist reliefs and inscriptions. Yuling, Qing Eastern Tombs

Eastern Cemetery. However, his final resting place is the Muling Mausoleum, which is part of the Western Tombs. Sources report that a tomb was begun for him in the Eastern Cemetery but the soil turned out to be inauspicious and humid, and the Emperor interpreted this as a sign that a change of location was required. As a matter of fact, I have identified a huge project for a royal tomb, which can still be perceived in aerial photographs. The project has cognitive characteristics that would tend to indicate an abandoned Duaogong project, rather than a possible original project by Yongzheng, or a planned tomb for Empresses or concubines. Firstly, it is located directly to the east of the second imperial complex, that of Kangxi, and it is therefore in the right place to be the *fourth* imperial complex, the third one being Xiaoling to the west side of the main axis. Secondly, it is oriented to the main peak to the south, and this is a characteristic which was apparently reserved for emperors' tombs.

Whatever the case, the successor of Duaogong, Xianfeng, respected the alternation and built his tomb—called Dingling—in the Eastern Necropolis. The place selected is to the west of the main axis, and in this way a formal east-west alternation was also respected (if we take into account the unfinished project to the east) as well as avoiding an area which had proved to be inauspicious. However, the distance from the center of the necropolis became relevant and the architects chose to make use of a different "vein" of the dragon to the north of the tomb, and of a different peak, located to the west of Changrui mountain, to align to the south. When Xianfeng died, his successor was his six-year-old son, Zaichun. He was enthroned as Tongzhi under the regency of Empresses Cixi and Ci'an. The twin tombs of these two distinguished women are located near the Dingling complex and called Puxiangyu East Dingling and Putuoyu East Dingling respectively. The setting of these tombs is particularly fascinating, and the tombs themselves boast refined decorations. As mentioned, the interior rooms of Cixi's tomb had been violated and are therefore visible. This is the only tomb of an empress to be excavated in China so far (Liu 2013) (Fig. 8.12).

Emperor Tongzhi did not observe the alternation rule and chose to be buried in a mausoleum, Huiling, located in the Eastern Cemetery. He did, however, respect local alternation so the architects had to find a place to the east of the main axis. Apparently no suitable place was found any closer than many kilometers to the south-east, in plain flatland. Tongzhi died without a son to succeed him; breaking the imperial convention, Cixi managed to nominate her nephew (the cousin of the Emperor) Zaitian, who was enthroned as Guangxu. He was only four, so that the two ladies remained in charge of the country up to 1881, when the Guangxu Emperor was nine and Empress Dowager Ci'an died unexpectedly, leaving Cixi as the sole Regent. Guangxu, whose tomb is in the Western cemetery, died on 14 November 1908, one day before Cixi's death. He was the last Emperor of China to build his tomb in a royal Necropolis. He was succeeded by his nephew, the child Puyi, who ascended as Emperor Xuantong. After three years, the Regent Empress Dowager Longyu signed the abdication decree on his behalf, putting a definitive end to imperial rule in China. Puyi later became the puppet Emperor of Manchuria during the Japanese invasion, and was put on trial for war crimes after the Second World War; having served his

Fig. 8.12   Replicating the landscape: the burial mound of Empress Cixi, Eastern Qing Tombs

sentence, he ended his life quietly in Beijing and died in 1967 (his story is told in a Bernardo Bertolucci's unforgettable movie).

8.3 One Dynasty, Two Necropolises

After having described the development of both necropolises, we shall now visit the Western Tombs in details. As mentioned, this necropolis was founded by Kangxi' son, Emperor Yongzheng, and was used by three other Emperors (Jiaqing, Daoguang and Guangxu) and their relatives, giving a total of fourteen mausoleums (Fig. 8.13).

The place is located in the Yi County area, north-west of Beijing; it is "protected" from the north by a mountain range, Yongning, and from the south by a river, the Yishui; an impressive, ancient forest of pines surrounds the whole site. However, the mountain peaks do not encircle the site, and the river is quite far south; in actual fact, the tombs are scattered around the foot of the hills in a virtually continuous way for several kilometers. We shall come back to this problem later on; for the moment, a better understanding of the necropolis can be obtained by following its architectural development, as we did for the Eastern Tombs.

As in Fengtaling, also here the first project, Yongzheng's tomb Tailing, included a very long and scenic Spirit Road. The path, as usual, is not completely straight, and bends encircling a small hill upon arrival at the tomb complex; however, the main section defines an axis oriented cardinally (azimuth 2°) and pointing to one

Fig. 8.13 The imperial tombs and their satellites in the Qing Western tombs area, numbered in chronological order: (1) Yongzheng (Tailing) (2) Jiaqing (Changling) (3) Daoguang (Muling) (4) Guangxu (Chongling). *Credits* Images courtesy Google Earth, editing by the author

among two low peaks located to the south. The architecture and organization of the complex are very similar to those of the eastern tombs, but a distinctive feature is the magnificent open hall bordered on all the four sides by stone arches (Figs. 8.14, 8.15 and 8.16).

The second imperial project, Changling of Emperor Jiaqing, was linked to the first one in a manner very similar to what happened for the second project in the Eastern tombs, by adding a branch to the Spirit Path. However, the complex is to the west of the first and does not point to the same peak, but to the one nearby. At this point, no other favorable place was available in the vicinity and, therefore,

Fig. 8.14 The main axis at the Western Qing tombs (north to the left): (1) Tailing (2) Bend in the Sacred Way (3) Main bridges (4) Stele pavilion (5) Four sided ceremonial gate (6) South Hill. *Credits* Images courtesy Google Earth, editing by the author

Fig. 8.15 Four-sided ceremonial gate, Qing Western Tombs

Fig. 8.16 Stone Elephant with elaborated harness, Qing Western Tombs

the third and the fourth imperial tombs complexes were built very far, more than 5 km off as the crow flies, to the west and to the east of the first, respectively. The morphology of the terrain did not allow for the construction of Spirit Paths connected to the first one, so the tombs lack this element. Of these two complexes, the first, Muling of Emperor Daoguang is unusual in many respects. He was probably the owner of the unfinished project we encountered in the Eastern Tombs, and when the terrain proved unsuitable there and the Emperor decided to move to the Western Necropolis, the scale and grandeur of the tomb were planned to be considerably smaller compared with the other two already present. However, the decorations and materials used are the finest possible. In particular, the Long'en Hall in Muling is in fine wood (nanmu) and decorated with dragon figures. Another distinctive feature is the orientation, which is not topographical. While the other royal mausoleums were oriented topographically with the choice of a relatively acceptable (from the geomantic point of view) peak to the south, Muling has an azimuth of only 32° south of east, and the horizon is nearly flat (Table A.8). The corresponding astronomical declination is impressively close to that of the sun at the winter solstice, which thus rises in alignment with the tomb. It may well be that the forced shift to the Western Tombs somehow impelled the Emperor also to choose a different kind of orientation, astronomical rather than geomantic.

   The last imperial tomb constructed in the Western Tombs area was also the last one built in China, Gaunxu's Chongling, located northeast of Yongzhen's one. Looted in 1938 and restored in 1980, it is the only tomb of the Western Necropolis that has ever been opened.

   All things considered, the Qing left us two fascinating, albeit quite different, sacred landscapes. The Eastern tombs took advantage of an almost unique Feng Shui alignment between the mountain to the north and the hill to the south, to produce a complex landscape whose main element is the sacred road originally built for the first complex. The main section of this road, the tombs of Qianlong and Tongzhi, and the unfinished imperial project are all oriented to the southern hill. Overall, the orientations of all the Qing Eastern Tombs (Table A.7) was plainly governed by geomantic topographical factors, as also confirmed by the correlation coefficient test versus the behavior of the magnetic declination (it returns a negative value $R = -0.71$ showing complete absence of correlation).

   The Western tombs, although located in a beautiful natural environment, do not exhibit any special features relating to Feng Shui rules of the landscape. Given the analysis we have carried out, perhaps a possible conclusion is that the place was not meant to become a necropolis: in the mind of its founder, it had to remain only the place of his—intentionally separate—tomb. Whatever the reason for this separation, I have tried to repeat (using satellite imagery) the task that the imperial geomancers were asked to solve, namely to find a suitable place roughly to the north of Beijing, excluding of course Fengtaling and the Ming Valley. The results are quite intriguing. Indeed, the overall geographical situation is as follows: Beijing lies in the northeast part of the so-called North China Plain, a gigantic alluvial plain bordered to the north by the Yanshan Mountains, to the west by the Taihang Mountains, to the south by the Dabie and Tianmu Mountains, and to the east by the Yellow Sea. The Yellow River flows through the middle of the plain. The arc of the Yanshan mountains is not continuous but has a vast "opening" to the north; it is exactly on the border of this opening that the Valley of the Ming Tombs is located. Looking right (east), Fengtaling stands out as the only place where the hills are sufficiently *not* smooth to assure a scenic front to the north; in addition, of course, it enjoys the spectacular protective mountain to the south. Looking left (west), on the other hand, one does encounter a place that any geomancer would deem suitable, enclosed as it is by dragon-shaped hills: the Fangshan area. However, in Qing times it was a rather overcrowded area, with many sacred places that had stood there for millennia. The next, and very last, place which offers some of the required characteristics is the chain of hills which unfolds for scores of kilometers to the west of Fangshan. So, there was no other choice than choosing a position there. How was this position selected? In my (digital) geomantic explorations, I have noticed that—either by chance or by design—the distance as the crow flies between the center of the Forbidden City in Beijing and the first mausoleum built in the eastern necropolis is 111.8 km, a value very close to the distance calculated from the same point and the first tomb built in the western necropolis, which is 108.4 km (the difference is less than 3%). Of course the two sites are not inter-visible, but precise geodetic measurements were certainly within the capability of the Chinese geographers of the time. This raises the possibility

Fig. 8.17 The positions of the Ming tombs (1) and of the Qing Western (2) and Qing Eastern (3) tombs with respect to the Forbidden City (F) in Beijing. *Credits* Images courtesy Google Earth, editing by the author

that, once it was decided to move from the Eastern Necropolis, the first of the Qing Western Tombs was built where the circumference centered on the Forbidden City, with radius equal to the distance from the Eastern tombs, intersects the border between the valley and the hills. Interestingly, taking into account the position of the Ming tombs, this creates a (roughly) symmetrical configuration (Fig. 8.17).

The location of the Western Qing tombs might thus have been selected more in accordance with principles of "Geomantic Geography", rather than for its local Feng Shui features.

Conclusions: A View from Purple Mountain

The sacred landscapes of imperial China we have come across in this book are outstanding examples of how deep and harmonious the relationship between the human-built environment and natural settings can be. In particular, there was a special, exclusive connection between the rulers' Mandate of Heaven and the landscape they chose for their residences and their burial sites. This led to the development of several "microcosm" landscapes where the energy and the beauty of the country was replicated and encapsulated in order for the royal dwellings to become an explicit icon of the ruler's power, and to the development of complex royal necropolises, each one with special characteristics, but all devoted to explicitly manifesting the divine rights of the dynasty.

The first Emperor's dream of immortality led him to choose a landscape dominated by Mount Li, in which his own huge, artificial mountain was added harmoniously: Shihuang's mound was literally "a copy of a mountain".

To erect mountains in a place where there were no mountains was, on the other hand, the choice of the Han rulers, who placed their mausoleums in the Wei floodplain; as we have seen, the fascinating, inter-connected landscape formed by their huge pyramids can still be appreciated today. Their orientation was astronomical, but their mutual placement followed a principle of alternation.

Real mountains—but provided with spectacular sacred paths oriented astronomically—were the solution for the Tang Emperors, while the Song opted for more modest complexes built in their homeland, in the shadow of the central sacred mountain of China. Also in this case, astronomy governed the orientation of the Spirit Paths, while the mutual placement followed a orderly rule.

During the Song, a complicated view of the relationship of landscape with human life and burial became predominant: the doctrine called Feng Shui. This doctrine developed along two lines: a theory of forms, based on the morphology of lands and the flow of waters, and a compass school, based on magnetic bearings measures. We found, however, no traces of the application of Compass Feng Shui either in royal estates or in royal tombs, while Form Feng Shui was extensively applied, for instance in the Ming Imperial Necropolis and in the Qing Eastern Necropolis. Consequently, I have tried to explore here the connection between a Feng Shui landscape and a sacred landscape in Eliade's sense: a place which is founded to be inhabited, and

G. Magli, *Sacred Landscapes of Imperial China*,
https://doi.org/10.1007/978-3-030-49324-0

Fig. C.1 Purple Mountain, Nanjing. The mausoleum of Dr. Sun Yat Sen at sunrise. *Credits* Images used under license from Shutterstock.com

an arena where hierophanies occur. In Feng Shui, the "object" landscape turns out to be a sort of living entity, which is endowed with features which can add up to a increasing coefficient of auspiciousness, or sacredness. In a sense, then, a Feng Shui landscape is a sacred space where a perennial, multi-composite hierophany takes place. A contribution to this hierophany is made by living nature—hills, waters, vegetation, illumination—as well as the built environment, provided—again—that it is constructed following specific rules of placement, orientation and so on. The result, as a matter of fact, seems to be culture-independent: for example, the intrinsic sense of beauty with which a place like the Valley of the Thirteen Tombs is imbued appears to be universally recognized.

An entire chapter of the history of architecture in China, however, that regarding urban planning, has still to be analyzed from the cognitive point of view. It is well-known that simple geometric and mathematical rules for the planning of the cities were codified during the Han, adding a treatise to an already existing ritual text, the Rites of Zhou. These rules dictated a square plan divided into 3 × 3 equal sectors (the so-called Magic Square), with three gates per side and oriented to the cardinal points. Many studies have investigated on how these rules were applied and on their possible merging with Feng Shui rules in later times (Schinz 1996; Wheatley 1971; 1975; Xu 2019); however, a complete and systematic approach to the orientation and siting of ancient Chinese towns is still missing, and I hope to carry out a complete investigation on this topic in the future. In the meantime, I think that this book should close with an homage to the millions of anonymous people who were involved, and in many cases forced, to participate in the gargantuan projects aimed at assuring the entertainment and the afterlife of the individuals we have met in this book.

To do this, let me take you again to Purple Mountain, Nanjing (Fig. C.1).

On January 1, 1912 the first President of China Dr. Sun Yat Sen proclaimed the birth of the Republic, changing forever the future of this magnificent country. On his death, occurring in 1925, Sen was awarded a huge mausoleum, built at walking distance from the first Ming Mausoleum, Xiaoling. The project develops along an ascending, linear axis towards one of the peaks of Purple Mountain and actually recalls many of the features of the imperial tombs. In particular, the huge marble gate is inspired by those built by the Ming six hundred years before.

As one climbs towards the summit, on the gate of the Mausoleum, a motto can still be read, which in my view is also the best possible conclusion for this book.

The motto says: *Tian Xia Wei Gong*, that is, *What is under heaven, is for everyone.*

Appendix
Probing Feng Shui Landscapes

One of the aims of this book is to develop a scientific approach to Feng Shui land-scapes. This, of course, does not mean probing the validity of Feng Shui ideas—for instance, the existence of Qi—for the simple reason that such ideas have no scientific validity. What we wish to do is to understand the beliefs and the way of thinking of the people (the architects, and those who commissioned their work) who conceived such landscapes and the buildings constructed in them. Thus, we need to develop as rigorous a method as possible to understand if the planners and the architects of a site were working with Feng Shui ideas in mind, and how these ideas were implemented. At first sight, this may appear simple, but this is not the case. Indeed, practically *any* ancient Chinese landscape is guaranteed to be a genuine Feng Shui landscape in all sort of publications, not just tourist ones. Of course, this cannot be true, and, indeed, it is not: the situation is almost paradoxical, as even the most world-famous Chinese site, the tomb of the first Emperor (described in Chap. 3) is sometimes declared to be an example of a Feng Shui landscape, while it is the exact opposite (mountains to the south, river to the north). In addition, Feng Shui itself is quite elastic, and different "masters" may consider specific features in different ways or with a differing order of importance (Yoon 2008; Tembata and Okazaki 2012). Examples even exist of important ancient places which were already said in texts of the time to be especially auspicious, but, when their characteristics are inspected in detail, turn out not to conform to well-established Feng Shui "master lines". What probably happened in these cases was, that the choice of the place was almost forced, and so the geomancers exalted some characteristics and tried to adapt the others in developing the building projects. However, generally speaking, the main Feng Shui characteristics of a place *are* recognizable and satellite imagery is of great help in doing this. As a matter of fact, it emerges that in (Form) Feng Shui-based landscapes, specific kinds of alignments—which we will hence call *Feng Shui alignments*—can be identified and studied with the same methods that Archaeoastronomy, the study of the presence of astronomy in the planning of ancient monuments, employs for astronomical alignments. These Feng Shui alignments are straight lines, visually connecting the axes of the buildings—and also, typically, other elements like the stelae pavilions—with relevant topographical features on the horizon at one or both ends. To study Feng Shui alignments, we shall thus use—as in Archaeoastronomy (see e.g. Magli 2020)—a series of tools: the azimuth-altitude reference system, the horizon formula, and a virtual globe software. Whereas to study the possible presence of Compass Feng Shui alignments—and therefore, alignments governed by magnetic directions—we shall use palaomagnetic data sets.

The Azimuth-Altitude Reference System

The azimuth-altitude reference system allows us to identify each point on the celestial sphere, the virtual spherical surface which represents the sky as seen from an observer on the Earth. This point can be for instance the Sun or a star at a certain moment of their path in the sky, but can be also a mountain peak or the summit of a temple seen from a fixed position. Since to fix a point on a sphere, only two numbers are needed (for instance, latitude and longitude on the Earth), the azimuth-altitude reference system associates two numerical values to any object of interest. To visualize the reference frame, imagine first prolonging the course of the Earth axis onto the celestial sphere: the ideal point of intersection is the celestial pole (north or south depending on the hemisphere of the observer; but in this book we are interested in the northern hemisphere). Now, we should imagine lowering the perpendicular to the horizon from the celestial pole. In this way, we can identify a point on the horizon and a direction on the ground pointing towards it: these are the geographical north and the meridian. Given a point S in the sky, whose coordinates we wish to find, we now imagine tracing the vertical plane which passes through this point. This plane intersects the horizon of the observer at a point, say S*; the azimuth is the angle between north and point S* on the horizon, counting positively from north to east (in other words, clockwise), and the altitude is the angle measured on the vertical circle from S* to S (Fig. A.1).

The Horizon Formula

We need to ensure that the alignments we eventually find do not occur by chance, and the first step in doing this is to check that they are connecting inter-visible objects. To do this we need to use a simple tool called the *horizon formula*. This formula originates in the fact, of course, that the Earth is round, so that two points can be claimed to have been intentionally aligned only if they are (or at least were, if modern changes in the landscape have interrupted the visual lines) inter-visible. Curiously enough, the roundness of the Earth is in many case forgotten: Internet in particular is full of websites claiming alleged "alignments" between places even hundreds of kilometers apart, which are drawn as straight lines on the spherical surface of our

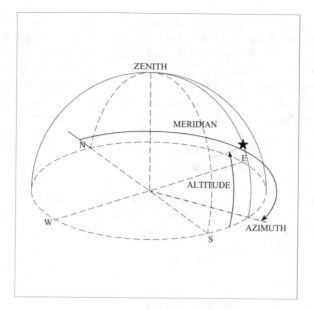

Fig. A.1 The altitude/azimuth coordinate system

Fig. A.2 The geometry
needed for the horizon
formula. Given a point P
located at vertical height h
with respect to the earth
surface P′, the farthermost
point W that can be seen
from P is the point where
passes the tangent to the
earth circumference from P.
The horizon formula is an
estimate of the distance d =
PW valid when h is small
with respect to the earth
radius

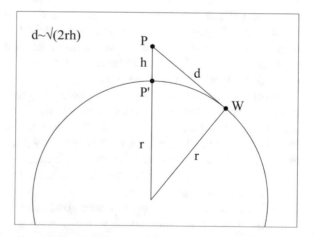

planet (as air pilots know very well, the only concept of a "straight line" between
two non-intervisible points that has any sense on the Earth's surface is the shortest
path between them, which is the arc of the maximal circle between the same points).
To distinguish truly visual—and thus possibly non-coincidental—alignments from
all the rest, one simply has to use the Pythagorean theorem (Fig. A.2).

Given a point P located at vertical height h with respect to the Earth surface, the
furthermost point W that can theoretically be seen from P is the point where it passes
the tangent to the Earth's circumference from P. Clearly then, the distance d between

P and W satisfies $d^2 = (r + h)^2 - r^2$, so that $d = \sqrt{(2rh + h^2)}$ where r is the Earth's radius. If the heights are very small with respect to r, we can safely discard h^2 (to see this, rewrite as $d = \sqrt{2rh(1 + h/2r)}$ and observe that h/2r is much smaller than 1), so that to all extent it is $d = \sqrt{(2rh)}$. Finally, since $2r \sim 13{,}000$ km, we conclude that the visible horizon d in kilometers equals, with a fair approximation, the square root of 13 h if h is expressed in meters. This is the "horizon formula"; somewhat surprisingly, it shows that the Earth's curvature (which, as long as we do not travel by plane for long distances, is seldom part of our daily experience) actually has a very significant effect: for instance, a person two meters tall has a visible horizon d $\sim \sqrt{(26)}$ kilometers, that is only slightly greater than 5 km. Very high objects built by humans have visibility in the order of tens of kilometers: for instance, a Maya pyramid 70 m tall (as is the so-called "Temple 4" at Tikal, Guatemala) is theoretically visible from a distance d $\sim \sqrt{(910)}$ kilometers, so from about 30 km afar. It should further be noted that the theoretical visibility distances can be significantly reduced by local atmospheric conditions.

The Use of Virtual Globe Software

Azimuth, altitudes and distances can be measured in the field with simple instruments, and of course a direct knowledge of the places under examination is always fundamental. In particular, I have personally visited the vast majority of the sites mentioned in the present book. However, there are digital instruments which are of enormous help in studying sacred landscapes: the so-called *Virtual Globe Softwares*, of which the most popular is Google Earth.

A virtual globe is essentially an extremely accurate computer version of a world globe. It maps out the Earth, area by area, by superimposing images obtained from satellite imagery. It also provides 3D reconstructions of buildings and street-view navigation. The resolution is, in most cases, very good and gives a fair picture of the sites. A virtual globe is very helpful in the cases where landscape and visibility were different in ancient times, since the program allows us to study visibility lines that are today lost or broken up by intervening obstacles. The program also allows—through the use of the ruler option—for a good estimate of azimuths and distances; if one requests the elevation profile of the lines of interest, the program visualizes the elevation between the two extremes of a chosen path and thus also permits the calculation of the altitude in degrees. On account of the high quality of the images and of the low projection error associated with them, the intrinsic error expected from Google Earth measures is quite low (Potere 2008; Luo et al. 2018; Guo 2013). For this reason, and for the sake of internal consistency and homogeneity, all data reported in the tables of this book have been obtained through satellite imagery. However, checks of sample data in each table have been done by other means, mostly with readings on site by the author but also by comparisons with existing surveys. The results of these tests confirm the validity of the satellite imagery approach.

Palaeomagnetic Models

Let us now consider the problem of developing a rigorous scientific test to verify the presence of *Compass* Feng Shui alignments, and accordingly the problem of understanding whether magnetic orientation was used in the planning of a site. To tackle this, we need to introduce a few notions of Geomagnetism.

We live on the solid surface of our planet, but this surface is only an external, very thin crust (from 5 to 70 km deep). To analyze what is below, volcano's samples, as well as indirect measurements like the study of seismic waves, are used. As a result, we know that the internal structure of the Earth can be thought of as a series of spherical layers; under the solid crust, we find a highly viscous mantle, a liquid outer core, and—probably—a solid, iron, innermost core. In particular, the liquid outer core appears to consist mostly of iron mixed with nickel; electric currents circulating in it are responsible—through a natural process called Geodynamo—for the Earth's magnetic field. It is now essential to understand that there is only a rough correspondence between the Earth's magnetic poles and the Earth's rotation poles and between the force lines of the Earth magnetic fields and the meridians. In addition, the magnetic poles *continuously move* over the Earth's surface. To measure the direction of the magnetic field we can use a magnetic compass: an instrument based on a magnetized pointer turning freely upon a pivot, in air or in a stabilizing fluid. When the compass is held level, the pointer turns until it stabilizes, pointing in a certain direction called *magnetic north*. The compass is equipped with a circular grade scale which allows for the determination of *magnetic azimuths,* that is angles measured with respect to the direction of the magnetic north. As mentioned, magnetic north does not correspond to geographical north (which of course is constant and fixed) and, moreover, varies continuously. The difference between magnetic north and geographical north at a given place and at a given time is called *magnetic declination*; it is assumed positive if it is to the east of north, negative otherwise (Fig. A.3).

The study of the time variation of the magnetic field is called Palaeomagnetism. In terms of thousands of years taken as time steps, Palaeomagnetism studies the reversals of the magnetic poles which are registered in rocks and sediment cores (the most recent one occurred around 780,000 years ago). However, in recent years, this discipline has made great advances in reconstructing the field values, not only in geological terms but also for historical times, which is the area of interest for us here. Information about the strength and direction of the magnetic field (prior to the establishment of geomagnetic scientific archives) can be obtained from lava flows, but also from archaeological materials of secure dating. As a consequence, fairly reliable *Palaeomagnetic models*, that is models of the variation in time of the magnetic declination, have been produced. In this book I have adopted the CALS10k.2 model, developed at the German Research Center for Geosciences at Potsdam (Korte et al. 2011; Constable et al. 2016). The model returns the expected magnetic declination at a given historical time and at given geographical coordinates, and is of fundamental importance for studying the history of Compass Feng Shui. Indeed, the variation in time of the magnetic declination has the—apparently paradoxical—consequence that *it is impossible to establish with a magnetic compass whether a monument was oriented with a magnetic compass*, since the azimuth of the magnetic north

Fig. A.3 The magnetic
declination

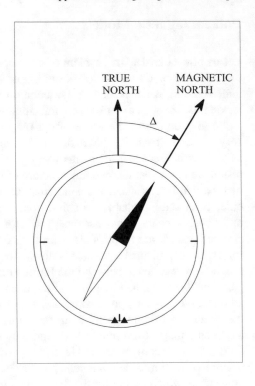

today is generally different—and it can be extremely different—from the one shown
by a magnetic compass at the time of construction. Therefore, to test for the use
of compass Feng Shui we proceed as follows. Consider, for instance, the axis of an
imperial tomb. The axis is first measured and recorded, together with the geographical
coordinates. Then a reference time is associated with the monument: the year when
the construction was started, if known, or, if not, the year of accession of the owner.
The variation of the magnetic declination within a few years is usually small (in
the order of some arc minutes per year) so that the procedure is almost independent
from the specific year chosen. Once position and date of interest are known, the
program calculates the magnetic declination at the given place and date, and so
returns the azimuth of the direction "north" which was shown by a compass *if used
by a geomancer at the site at the moment of its planning*. Of course, if the instrument
was effectively used to align to the north, this azimuth must coincide with the *real*
(not the magnetic) azimuth as measured today. The procedure is then repeated for all
the monuments of a site. Once all the palaeomagnetic data and the orientation data
have been collected, the two data sets can be compared using a test of correlation.
Fortunately, if a correlation exists it must necessarily be a linear and positive one, so
that it is sufficient to use a simple mathematical tool (the so-called Pearson correlation
test R) and check if the value obtained—which is always between -1 and $+1$—is
close to $+1$; if not, there is no correlation (for details on the test see for example
James et al. 2013).

Chronology of Imperial China

Notice: only the dynasties dealt with in details in the book are expanded as full lists of rulers with reign years.

Xia Dynasty 2100–1600 BC
Shang Dynasty 1600–1050 BC
Zhou Dynasty 1050–256 BC
Spring and Autumn Period 770–475 BC
Warring States Period 475–221 BC

Qin Dynasty 221–206 BC
Shihuang 221–210 BC
Er Shi 210–206 BC

Western Han Dynasty 202 BC–9 AD
Gaozu 202–195 BC
Hui 195–188 BC
Wen 180–157 BC
Jing 157–141 BC
Wu 141–87 BC
Zhao 87–74 BC
Xuan 74–49 BC
Yuan 49–33 BC
Cheng 33–7 BC
Ai 7–1 BC
Ping 1 BC–6 AD
Ruzi 6–9 AD

© The Editor(s) (if applicable) and The Author(s), under exclusive license
to Springer Nature Switzerland AG 2020
G. Magli, *Sacred Landscapes of Imperial China*,
https://doi.org/10.1007/978-3-030-49324-0

Wang Mang 9–23 AD (Xin dynasty)
Eastern Han Dynasty 25–220 AD

Guangwu 25–57 AD
Ming 57–75 AD
Zhang 75–88 AD
He 88–106 AD
Shang 106 AD
An 106–125 AD
Shun 125–144 AD
Chong 144–145 AD
Zhi 145–146 AD
Huan 146–168 AD
Ling 168–189 AD
Xian 189–220 AD
Three Kingdoms 220–280 AD
Jin dynasty 266–420 AD
Northern and Southern dynasties 420–581 AD
Sui Dynasty 581–618 AD

Tang Dynasty 618–906 AD
Gaozu 618–626 AD
Taizong 626–650 AD
Gaozong 649–683 AD
Zhongzong 684, 705–710 AD
Ruizong 684–690 AD
Wu Zetian (Zhao dynasty) 690–705 AD
Xuanzong 712–756 AD
Suzong 756–762 AD
Daizong 762–779 AD
Dezong 779–805 AD
Shunzong 805 AD
Xianzong 805–820 AD
Muzong 820–824 AD
Jingzong 824–827 AD
Wenzong 826–840 AD
Wuzong 840–846 AD
Xuanzong 846–859 AD
Yizong 859–873 AD
Xizong 873–888 AD
Zhaozong 888–904 AD
Zhaoxuan 904–906 AD

Five Dynasties 907–960 AD
Northern Song dynasty 960–1127 AD
Taizu 960–976 AD

Taizong 976–997 AD
Zhenzong 998–1022 AD
Renzong 1023–1063 AD
Yingzong 1064–1067 AD
Shenzong 1067–1085 AD
Zhezong 1086–1100 AD
Huizong 1101–1125 AD
Qinzong 1126–1127 AD

Southern Song dynasty 1127–1279 AD
Western Xia dynasty 1038–1227 AD
Yuan Dynasty 1271–1368 AD
Ming Dynasty 1368–1644 AD

Hongwu 1368–1398 AD
Jianwen 1399–1402 AD
Yongle 1403–1424 AD
Hongxi 1425 AD
Xuande 1426–1435 AD
Yingzong 1435–1449/1457–1464 AD
Jingtai 1450–1456 AD
Xianzong 1465–1487 AD
Hongzhi 1488–1505 AD
Zhengde 1506–1521 AD
Jiajing 1522–1566 AD
Longqing 1567–1572 AD
Wanli 1573–1620 AD
Taichang 1620 AD
Dianqi 1621–1627 AD
Chongzheng 1628–1644 AD
Qing Dynasty 1644–1911 AD

Shunzi 1644–1661 AD
Kangxi 1662–1722 AD
Yongzheng 1723–1735 AD
Qianlong 1736–1795 AD
Jiaqing 1796–1820 AD
Daoguang 1821–1850 AD
Xianfeng 1851–1861 AD
Dongzi 1862–1874 AD
Guangxu 1875–1908 AD
Puy (Xuantong) 1909–1911 AD

Tables

See Tables A.1, A.2, A.3, A.4, A.5, A.6, A.7 and A.8.

Table A.1 Mausoleums of the Western Han Dynasty

	Emperor	Associated burials	Azimuth of the "eastern" side looking north	Name of tomb	Notes	Mag. Dec
1	Gaozu 202–195 BC		347	Changling		−1.54
		Empress Lü	347		SE of main	
2	Hui 195–188 BC		346	Anling		−1.22
		Empress Zhang	347		NW of main	
		Marquis Zhang	346		NE of main	
		Princess Lu	345		NE of main	
W	Wen 180–157 BC			Baling	Natural hill	
		Empress Dou	294		Towards Baling at 24	
		Empress Bo	294			
3	Jing 157–141 BC		359	Yanling		1.31
		Empress Wang	359		NE of main	
4	Wu 141–87 BC		352	Maoling		2.19
		Empress Li	352		NW of main	
5	Zhao 87–74 BC		351	Pingling		5.07
		Empress Shanguan	351		NW of main	
6	Xuan 74–49 BC		360	Duling	Opposite river of Wei	
		Empress Wang	360		SE of main	
		Empress Xu	360		SE of main	
7	Yuan 49–33 BC		359	Weiling		6.43
		Empress Wang	359		NW of main	
8	Cheng 33–7 BC		350	Yangling		

(continued)

Table A.1 (continued)

Emperor		Associated burials	Azimuth of the "eastern" side looking north	Name of tomb	Notes	Mag. Dec
		Empress Xu	351		With 7 other small mounds	
		Consort Ban	350		NE of main	
9	Ai 7–1 BC		360	Yiling		7.54
		Empress Fu	357		NE of main	
10	Ping 9 BC–AD 6		360	Kanling		7.93

Table A.2 Mausoleums of the Tang Dynasty

Emperor	Name of tomb	Date of Accession	Azimuth (looking towards the mountain)	Magnetic declination	Notes	
Gaozu	Xianling	618	359	1.32	Mound	1
Taizong	Zhaoling	626	190	1.32	Spirit Path from the north	2
Gao Zong	Qianling	649	354	1.16		3
Zhongzong	Dingling	705		−0.12	Not measurable	4
Ruizong	Qianling	684	356	0.49		5
Xuanzong	Tailing	712	348	−0.29	Topographical	6
Suzong	Jianling	756	359 and 348	−2.18	Topographical	7
Daizong	Yuanling	762		−2.41	Not measurable	8
Dezong	Chongling	779		−3.34	Not measurable	9
Shunzong	Fengling	805	350	−4.45		10
Xianzong	Jingling	805	352	−4.45		11
Muzong	Guangling	820	353	−5.05		12
Jingzong	Zhuangling	824	356	−5.23		13
Wenzong	Zhangling	827	355	−5.23		14
Wuzong	Ruiling	840	355	−5.72		15
Xuanzong	Zhenling	846		−5.86	Not measurable	16
Yizong	Jianling	859	358	−6.22		17
Xizong	Jingling	873		−6.48	Not measurable	18

Table A.3 Mausoleums of the Song Dynasty

	Emperor	Azimuth (spirit road towards mound)	Name of tomb		Magnetic declination at date of death
1	Xuanzu	Not meas.	Yongan		
2	Taizu 960–976	3	Yongchang		−5.6
3	Taizong 976–997	5	Yongxi		−4.6
4	Zhenzong 997–1022	4	Yongding		−3.6
5	Renzong 1022–1063	4	Yongzhao		−1.7
6	Yingzong 1063–1067	4	Yonghou		−1.5
7	Shenzong 1067–1085	1	Yongyu	Final section bends to 358°	−0.8
8	Zhezong 1085–1100	0	Yongtai		−0.1

Table A.4 Mausoleums of the Western Xia Dynasty

	Azimuth (enclosure, looking towards the mound)	Attribution	Magnetic Declination	Notes
1	0		−3.15	Probably constructed together with n.2 in 1038 AD
2	0		−3.15	
3	326			
4	355			
5	353			
6	348			
7	347	Renzong 1139–1193 AD	0.64	
8	Not meas.			
9	357			
U	345			Unfinished

Table A.5 Mausoleums of the Ming dynasty

Emperor	Name of Tomb	Accession	Az.	Mag. Dec.	Notes
Ming Ancestors	Zuling	1368	2	0.32	Huaian; built by Hongwu
Hongwu	Xiaoling	1368	0°	0.26	Nanjing
Yongle	Changling	1402	6	−0.36	
Hongxi	Xianling	1424	16	−0.83	
Xuande	Jingling	1425	51	−0.83	
Zhengtong	Yuling	1435	17	−1.00	
Chenghua	Maoling	1464	11	−1.42	
Hongzhi	Tailing	1487	346	−1.58	
Zhengde	Kangling	1505	292	−1.62	
Jiajing	Yongling	1521	50	−1.58	
Longqing	Zhaoling	1567	320	−1.25	
Wanli	Dingling	1572	299	−1.22	
Taichang	Qingling	1620	14	−1.26	
Tianqi	Deling	1620	84	−1.26	
Chongzhen	Siling	1627	358	−1.35	
Gongruixian (Jiajing's father)	Xianling	1521	31	−1.62	Zhongxiang

Table A.6 Early Qing tombs

Emperor	Name of tomb	Date	Az.	Mag. Dec.	Notes
Parents of Nurhaci	Dongjing	1624	324 and 64	−0.80	Liaoyang
Ancestor's of Qing dynasty	Yongling	1598	318	−0.70	Fushun
Nurhaci	Fuling	1616	344	−0.75	Shenyang
Hong Taji	Zhaoling	1626	357	−0.80	Shenyang

Table A.7 Qing Eastern Tombs (all Qing emperors are listed, also those buried in the western tombs)

	Emperor	Name of tomb	Accession date	Azimuth	Mag. Dec.	Notes
1	Shunzhi	Xiaoling	1650	334	−1.05	
1A				332		Wife
1B				345		Mother
2	Kangxi	Jingling	1661	354	−1.19	
2A				349		
2B				354		
	Yongzheng					Western tombs
3	Qianlong	Yuling	1735	329	−1.22	
3A				345		
	Jaqing					Western tombs
	Daoguang					Western tombs
4	Xianfeng	Dingling	1850	348.	−2.41	
4A				358		
4B				336		Twin tomb for Empress Dowager Cixi and Empress Dowager Cian
5	Tongzhi	Huiling	1861	9	−2.64	
5A				5		
	Guangxu					Western tombs
6				352		Abandoned construction site (Daoguang?)
6A				4		Two sons and two daughters of Daoguang

Table A.8 Qing Western tombs

	Emperor	Name of tomb	Date	Azimuth	Mag. Dec.	Notes
1	Yongzheng	Tailing	1722	347	−1.34	
1A				14		21 concubines
1B				9		
2	Jiaqing	Changling	1796	354	−1.31	
2A				352		17 concubines
2B				17		
2C				13		8th son
3	Daoguang	Muling	1820	302	−1.33	To winter solstice sunrise at 122
3A				7		
4	Guangxu	Chongling	1875	335	−2.15	
4A				20		Two concubines
4B				11		
4C				12		
4D				352		

References

Aguayo P, García-Sanjuán L (2002) The megalithic phenomenon in Andalusia (Spain): an overview. In: Proceedings of the colloquium origin and development of the megalithic phenomenon in Western Europe (Bougon, Paris), pp 451–476

Bai Y (2015) The studies on the measuring devices of the Han Dynasty and the relevant issues. Chin Archaeol 15(1):188–194

Bennet S (1978) Patterns of the Sky and Earth: a Chinese science of applied cosmology. Chin Sci 3:1–26

Berger PA (2003) Empire of emptiness: Buddhist Art and Political Authority in Qing China. University of Hawaii Press

Bigoni F, Bigoni D, Misseroni D, Wang D (2017) Megalithic stone beam bridges of ancient China reach the limits of strength and challenge size effect in granite. J Cult Her 26:167–171

Bonnet-Bidaud J, Praderie F, Whitfield S (2009) The Dunhuang Chinese sky: a comprehensive study of the oldest known star Atlas. J Astron Hist Herit 12:39–59

Brashier K (2011) Ancestral memory in Early China. Harvard-Yenching Institute Monograph Series 72

Braswell G (2012) Rain and fertility rituals in Postclassic Yucatan Featuring Chaak and Chak Chel. in The Ancient Maya of Mexico. Reinterpreting the Past of the Northern Maya Lowlands. Routledge, London

Brook T (1989) Ritual and the Building of Lineages in Late Imperial China. Harv J Asiat Stud 49(2):465–499

Bruun O (2008) An introduction to Feng Shui. Cambridge University Press, Cambridge

Buck DD (1975) Three Han Dynasty Tombs at Ma-Wangdui. World Archaeol 7(1):30–45

Bulling A (1966) Three popular motives in the Art of the Eastern Han period: the lifting of the tripod. The Crossing of a Bridge. Divinities. Archives of Asian Art 20:25–53

Burman (2018) The Terracotta Warriors: exploring the most intriguing puzzle in Chinese History. Pegasus, London

Campbell WH (2001) Earth magnetism: a guided tour through magnetic fields. Academic Press

Chang Y, Li T (1985) Application of mercury survey technique over the Mausoleum of Emperor Qin Shi Huang. J Geochem Expl 23(1):61–69

Charvátová I, Klokočník J, Kostelecký J, Kolmaš J (2011) Chinese tombs oriented by a compass: evidence from paleomagnetic changes versus the age of tombs. Stud Geophys Geod 55:159–174

Chongwen G (2007) The evolution of funerary ritual from the Pre-Qin to the Han Era. Chin Archaeol 7(1)

G. Magli, *Sacred Landscapes of Imperial China*,
https://doi.org/10.1007/978-3-030-49324-0

Coggins C (2014) When the land is excellent: Village Feng Shui Forests and the Nature of Lineage, Polity and Vitality in Southern China. In: Miller J, Smyer Yü D, van der Veer P (eds) Religious diversity and ecological sustainability in China. Routledge, Abingdon, pp 97–126

Coggins C, Chevrier J, Dwyer M, Longway L, Xu M, Tiso P, Li Zhen (2012) Village Fengshui Forests of Southern China–culture History and Conservation Status. ASIA Network Exchange 19(2):52–67

Coggins C, Minor J, Chen B, Zhang Y, Tiso P, Gultekin Cem (2019) China's Community Fengshui Forests—spiritual ecology and nature conservation. In: Verschuuren B, Brown S (eds) Cultural and spiritual significance of nature in protected areas. Routledge, New York, pp 227–239

Constable C, Korte M, Panovska S (2016) Persistent high paleosecular variation activity in southern hemisphere for at least 10,000 years. Earth Planet Sci Lett 453:78–86

De Boer H (2000) The geological origins of the oracle at Delphi, Greece. J Z Geol Soc Lond Spec Publ 171:399–412

De Groot JJM (1892) The religious system of China: its ancient forms, evolution, history and present aspect. Manners, Customs and Social Institutions Connected Therewith, 6 vols. E. J. Brill, Leyden

Didier JC (2009) In and outside the square: the Sky and the Power of Belief in Ancient China and the World, c. 4500 BC–AD 200. Sino-Platonic Papers 192

Drake FS (1943) Sculptured stones of the Han dynasty. Monumenta Serica 8:280–318

Eckfeld T (2005) Imperial Tombs in Tang China, 618–907: the Politics of Paradise. Routledge, London

Eitel E (1908) Feng Shui, or the Rudiments of Natural Science in China. Trubner & Co., Hong Kong

Eliade M (1959) The sacred and the profane: the nature of religion. Harcourt, London

Eliade M (1964) Shamanism: archaic techniques of ecstasy. Princeton, Princeton

Eliade M (1971) The myth of the eternal return: or, cosmos and history. Bollingen, London

Eno R (1990) The Confucian creation of heaven: philosophy and the defense of Ritual Mastery. State University of New York Press, New York

Fairbank W (1941) The offering shrines of Wu Liang Tzu. Harv J Asiat Stud 6(1):1–36

Ferguson JC (1931) The Six Horses at the Tomb of the Emperor T'ai Tsung of the T'ang Dynasty. Eastern Art 3:61–72

Field L (2001) The Zangshu, or Book of Burial by Guo Pu (276–324)

Fields LB (1989) The Ch'in Dynasty: Legalism and Confucianism'. J Asian Hist 23(1):1–25

Fitzgerald CP (1933) Son of heaven: a biography of Li Shih-Min, founder of the T'ang dynasty. Cambridge University Press

Flannery K, Marcus J (1996) Cognitive archaeology. In: Preucel RW, Hodder I (eds) Contemporary archaeology in theory: a reader (social archaeology). Wiley-Blackwell, New York

Fong MH (1984) Tang Tomb Murals Reviewed in the Light of Tang Texts on Painting. Artibus Asiae 45(1):35–72

Foret P (2000) Mapping Chengde: the Qing Landscape Enterprise. University of Hawaii Press, Honolulu

Forte M (2010) Western Han landscape and remote sensing applications at Xi'an (China). In: Campana S, Forte M, Liuzza C (eds) Space, time, place: third international conference on remote sensing in archaeology. Archaeopress, Oxford

Gong Q, Koch A (2002) Das Xiaoling Mausoleum des Kaisers Ruizong (662–716) im Kreis Pucheng. (Prov. Shaanxi, VR China) Monographien des RGZM 47, Meinz

Gowland W (1897) The Dolmens and Burial Mounds in Japan. Archaeologia 55(2):439–524

Guo Qinghua (2004) Tomb architecture of dynastic China: old and new questions. Architect Hist 47:1–24

Guo H (2013) Atlas of remote sensing for World Heritage: China. Springer, Berlin

Guolong L (2005) Death and the Otherworldly Journey in Early China as Seen through Tomb Texts, Travel Paraphernalia, and Road Rituals. Asia Major 18(1):1–44

Hannah R, Magli G (2011) The role of the sun in the Pantheon design and meaning. Numen-Arch Hist Relig 58(4):486–513

Hargett JM (1988) Huizong's magic marchmount: the Genyue pleasure park of Kaifeng. Monumenta Serica 38:1–48

Harley JB, Woodward D (1995) The history of cartography: cartography in the traditional east and Southeast Asian Societies. The University of Chicago Press

Harper D (1978) The Han Cosmic Board. Early China 4:1–10

He N (2018) Taosi: An archaeological example of urbanization as a political center in prehistoric China. Archaeol Res Asia 14:20–32

Hong J (2013) Exorcism from the Streets to the Tomb: an image of the Judge and Minions in the Xuanhua Liao Tomb No. 7. Arch Asian Art 63(1):1–25

Hotaling SJ (1978) The City walls of Han Chang'an. Toung Pao 64:1–46

Howard AF (2006) Chinese sculpture. Yale University and Foreign Languages Press, New Haven

Hsu M-L (1978) The Han Maps and Early Chinese Cartography. Ann Assoc Am Geograph 68(1):45–60

Hsueh-Man S (2005) Body Matters: Manikin Burials in the Liao Tombs of Xuanhua. Hebei Province. Artibus Asiae 65(1):99–141

Huo W (2008) Large-sized Stone-sculptured Animals of the Eastern Han Period in Sichuan and the Southern Silk Road. Kaogu 11:71–80

James JM (1988) The iconographic program of the Wu Family Offering Shrines (A.D. 151–ca. 170). Artibus Asiae 49(1):39–72

Jiao N (2007) On the Construction Concept of Western Han Dynasty Mausoleums. Archaeology 2007:78–87

Jiao N (2013) Shape factor analysis and inference of Western Han Dynasty Mausoleums. Archaeol Herit 2013:72–81

Jie S (2015) The hidden level in space and time: the vertical shaft in the royal tombs of the zhongshan kingdom in late Eastern Zhou (475–221 bce) China. Mater Relig: J Objects, Art Belief 11(1):76–102

Keightley DN (1991) The quest for eternity in Ancient China: the dead, their gifts, their names. In: Kuwayama G (ed) Ancient mortuary traditions of China: Papers on Chinese Ceramic Funerary Sculptures. Los Angeles County Museum of Art, Los Angeles

Keightley DN (1997) Graphs, words, and meanings: three reference works for Shang Oracle-bone Studies, with an Excursus on the Religious Role of the Day or Sun. J Am Orient Soc 117(3):507–524

Keightley DN (1998) Shamanism, death, and the ancestors: religious mediation in Neolithic and Shang China (ca. 5000–1000 B.C.) Asiat Stud 52:763–831

Keightley DN (2000) The ancestral landscape: time, space, and community in Late Shang China (ca. 1200–1045 B.C.). China Research Monograph 53 Institute of East Asian Studies, Berkeley

Kenderdine S, Forte M, Camporesi C (2005) Rhizome of Western Han: an omnispatial theatre for archaeology. In: Zhou M, Romanowska I, Wu Z, Kern M (eds) Revive the past. Computer applications and quantitative methods in archaeology (CAA), Proceedings of the 39th international conference, Beijing, 12–16 Apr. Text and Ritual in Early China, Seattle: University of Washington Press

Komlos J (2003) The size of the Chinese Terracotta Warriors. Antiquity 77

Korte M, Constable C, Donadini F (2011) CALS10k. 1: a holocene geomagnetic field model based on archeo- and paleomagnetic data. In: 25th IUGG General Assembly, Melbourne, Australia

Lewis R (2006) The construction of space in Early China. State University of New York Press, New York

Lin J, Armour for the afterlife. In: Portal, the first emperor, pp 181–189

Liu Y (2012) Chinas first emperor, the terracotta army and the Qin Culture. Minneapolis Institute of Arts, Minneapolis

Liu Y (2013) The categorization and periodization of the mausoleums of Qing Dynasty Imperial Consorts. J Gugong Stud 1

Liu L (2014) Qin Shi Huangs Mausoleum and Ancient Chinese Civilization. Ke Xue Chu, Beijing

Liu Y (2015) Beyond the First Emperor's Mausoleum: new perspectives on Qin Art. University of Washington Press, Seattle

Liu H, Zhou G, Wennersten R, Frostell B (2014) Analysis of sustainable urban development approaches in China. Habit Int 41:24–32

Loewe M (1979) Ways to paradise: the Chinese Quest for Immortality. Allen & Unwin, London

Loewe M (1985) The Royal Tombs of Zhongshan (c. 310 B.C.). Arts Asiatiques 40:130–134

Loewe M (2016) Problems of Han administration: ancestral rites, weights and measures, and the means of protest. Brill, London

Loewe M, Shaughnessy EL (eds) (1999) The Cambridge History of Ancient China: from the Origins of Civilization to 221 B.C. Cambridge University Press, Cambridge

Luo B, Grydeho'j A (2017) Sacred islands and island symbolism in Ancient and Imperial China: an exercise in decolonial island studies. Island Stud J 12(2):25–44

Luo L, Wang X, Guo H, Lasaponara R, Shi P, Bachagha N, Yao Y, Masini N, Chen F, Ji W (2018) Google Earth as a powerful tool for archaeological and cultural heritage applications: a review. Remote Sens 10:1558

Magli G (2012) The Snefru projects and the topography of funerary landscapes during the 12th Egyptian dynasty. Time Mind 5(1):53–72

Magli G (2013) Architecture, astronomy and sacred landscape in Ancient Egypt. Cambridge University Press, Cambridge

Magli G (2016) The Giza 'written' landscape and the double project of King Khufu. Time Mind 9:57–74

Magli G (2018) Royal mausoleums of the western Han and of the Song Chinese dynasties: a satellite imagery analysis. Arch Res Asia 15:45–54

Magli G (2019) The sacred landscape of the "Pyramids" of the Han Emperors: a cognitive approach to sustainability. Sustainability 11(3):789

Magli G (2019) Astronomy and Feng Shui in the projects of the Tang, Ming and Qing royal mausoleums: a satellite imagery approach. Arch Res Asia 17:98–108

Magli G (2020) Archaeoastronomy: introduction to the science of stars and stones. Springer, New York

Man J (2007) The Terracotta Army: Chinas First Emperor and the Birth of a Nation. Bantam Press, London

Martin-Torres M (2011) Making Weapons for the Terracotta Army. Archaeol Int 13:70

Meyer JF (1978) Feng-Shui of the Chinese City. Hist Relig 18(2):138–155

Meyer JF (1991) The Dragons of Tiananmen: Beijing as a sacred City, Univ of South Carolina Press

Miller AR (2015) Emperor Wen's 'Baling' mountain tomb: innovation in political rhetoric and necropolis design in early China. Paleogeogr Early Chin Hist 28:1–37

Miller J (2016) China landscapes, cultures, ecologies, religions. In: Jenkins JW, Tucker ME, Willis JG (eds) Routledge handbook of religion and ecology. Roulegde, New York

Nanfeng, J (2013) The analyses and identification of morphological components of the Western Han Mausoleums. Archaeol Cult Relics 5

Needham J (1956) Science and civilisation in China, vol 2: history of scientific thought. Cambridge University Press, Cambridge

Needham J (1959) Science and civilisation in China, vol 3: mathematics and the science of the heavens and the Earth. Cambridge University Press, Cambridge

Needham J (1965) Science and civilisation in China, vol 4: physics and physical technology, Part 1. Cambridge University Press, Cambridge

Needham J (1970) Clerks and Craftsmen in China and the West: lectures and addresses on the history of science and technology. Cambridge University Press, Cambridge

Nickel, L, The Terracotta Army, in Portal, The First Emperor, pp 159–179

Paludan A (1981) The imperial ming tombs. Yale University Press, Yale

Paludan A (1991) The Chinese spirit road: the classical tradition of stone tomb statuary. Yale University Press, Yale

Paludan A (1998) Chronicle of the Chinese Emperors: the reign-by-reign record of the rulers of Imperial China. Thames & Hudson Ltd., London

Pankenier D (1995) The Cosmo-Political Background of Heaven's Mandate. Early China 20:121–176

Pankenier D (2004) A brief history of Beiji (Northern Culmen): with an excursus on the origin of the character. J Am Orient Soc 124(2):1–26

Pankenier D (2009) Locating True North in Ancient China. Cosmology across cultures: Astronomical society of the pacific conference series, vol 409, pp 128–137

Pankenier D (2011) The cosmic center in early China and its archaic resonances. In: Ruggles CN (ed) Proceedings IAU symposium no. 278. Cambridge University Press, Cambridge, pp 298–307

Pankenier D (2013) Astrology and cosmology in Early China: conforming Earth to Heaven. Cambridge University Press, Cambridge

Paton J (2013) Five classics of Fengshui: Chinese spiritual geography in historical and environmental perspective. Brill Academic Pub

Ping X (2016) All the way to the Altair and the fable of cowherd and the weaving maiden. In: 2nd international conference on education technology, Management and Humanities Science ETMHS

Pirazzoli-t'Serstevens M (2009) Death and the dead: practices and images in the Qin and Han. In: Lagerwey J, Kalinowski M (eds) Early Chinese Religion I. Brill, London

Poo MC (2011) Preparation for the afterlife in Ancient China. In: Olberding Amy, Ivanhoe Philip (eds) Mortality in traditional Chinese Thought. State University of New York Press, Albany, pp 13–36

Portal J (2007) The first emperor: China Terracotta Army. Harvard University Press,

Potere D (2008) Horizontal positional accuracy of Google Earth's high-resolution imagery archive. Sensors 8:7973–7981

Powers M (1991) Art and political expression in Early China. Yale University Press, New Haven

Puett MJ (2002) To become a god: cosmology, sacrifice and self-divinization in Early China. Harvard University Press

Qingquan L (2010) Some aspects of time and space as seen in Liao-dynasty Tombs in Xuanhua. J Art Transl 2:1

Qingzhu L (2007) Archaeological discovery and research into the layout of the Palaces and Ancestral Shrines of Han Dynasty Chang'an—a comparative essay on the capital cities of Ancient Chinese Kingdoms and Empires. Early China 31:111–143

Rapoport A (1982) The meaning of the built environment. A nonverbal communication approach. University of Arizona Press, Tucson

Rawson J (1999) The Eternal Palaces of the Western Han. Artibus Asiae 59(1/2):5–58

Rawson J (2002) The power of images: the model universe of the first emperor and its legacy. Hist Res 75(188):123–154

Rawson J, The first emperors tomb: The Afterlife Universe, in Portal, The First Emperor, pp 114–145

Romain WF (2017) The archaeoastronomy and Feng Shui of Xanadu: Kublai Khan's imperial Mongolian capital. Time Mind 10(2):145–174

Romain WF (2019) Geomantic entanglements with mountains in Central Tibet: Royal Tombs of the Chongye Valley. European Association of Archaeologists, Bern

Schafer E (1968) Hunting parks and animal enclosures in Ancient China. J Econ Soc Hist Orient 11(3):318–343

Schafer E (1977) Pacing the void: T'ang approaches to the stars. University of California Press

Schinz A (1996) The magic square: history of Chinese City Planning. Edition Axel Menges

Segal EM (2009) Archaeology and cognitive science. In: Renfrew C, Zubrow EBW (eds) The ancient mind. Elements of cognitive archaeology. Cambridge University Press, Cambridge

Segalen V (1912) Les origines de la grande statuaire en Chine

Sivin N (1995) State, cosmos, and body in the last three centuries B. C. Harv J Asiat Stud 55(1):5–37

Smith RJ (2012) An overview of divination in China from the Song through the Qing: some issues and approaches. In: Divinatory traditions in East Asia: historical, comparative and transnational perspectives. Conference at Rice University, Houston

Steinhardt N (1993) The Tangut Royal Tombs near Yinchuan. Muqarnas, vol 10. Essays in Honor of Oleg Grabar, pp 369–381

Steinhardt N (2002) Chinese architecture. Yale University Press, New Haven

Stephenson FR (1994) Chinese and Korean Star Maps and Catalogs. In: Harley JB, Woodward D (eds) The history of cartography, vol 2. Cartography in the Traditional East and Southeast Asian Societies. The University of Chicago Press, Chicago, pp 511–578

Suhadolnik NV (2011) Han Mural Tombs: reflection of correlative cosmology through mural paintings. Asian Afr Stud XV(1):19–48

Sun L (2012) 2012 report on archaeology of the Tomb of Qin Shihuang 2009–2010. China Scientific Research Institute, Beijing

Swart P, Till BD (1984) Bronze Carriages from the Tomb of China's First Emperor. Archaeology 37(6):18–25

Tembata H, Okazaki S (2012) Relationships between Feng-Shui and landscapes of Changan and Heijo-Kyo. In: Archi-Cultural Translations Through the Silk Road, Nishinomiya, Japan

Thote A (2009) Shang and Zhou Funeral Practices: Interpretation of Material Vestiges. In: In: Lagerwey J, Kalinowski M (eds) Early Chinese Religion, Part 1, Shang through Han (1250 BC–220 AD), vol 1. Brill, Leiden, pp 103–142

Turnbull S (2009) Chinese Walled Cities 221 BC-AD 1644. Osprey Publishing

Wakeman F (1985) The great enterprise: the Manchu Reconstruction of Imperial Order in Seventeenth-Century China. University of California Press, Berkeley

Waldron A (1990) The great wall of China: from history to myth. Cambridge University Press, Cambridge

Wan (2009) Building an Immortal Land: The Ming Jiajing Emperor's West Park Asia Major 22(2009):65–99

Wang E (2011) Ascend to heaven or stay in the tomb? Paintings in Mawangdui Tomb 1 and the Virtual Ritual of Revival in Second-Century B.C.E. China. In: Olberding A, Ivanhoe J (eds) Mortality in traditional Chinese thought. New York State University Press, New York

Wang E (2012) What happened to the first emperor afterlife Spirit? In: Yang L (ed) Chinas First Emperor, pp 210–227

Wang E (2015) Afterlife entertainment? The Cauldron and Bare-torso Figures at the First Emperor's Tomb, in Liu Yang, Beyond the First Emperor's Mausoleum, pp 59–95

Wang D (2015) The archaeological investigation and coring exploration of the Weiling Mausoleum. Chin Archaeol 15:144–151

Wang F (2016) Geo-architecture and landscape in China's Geographic and Historic Context Volume 1 Geo-Architecture Wandering in the Landscape. Springer

Wang F, De Luca L (2018) When script engravings establish a new spatial dimension in a monument: the Tomb of Manchu Emperor Qianlong (18th century) International Preservation News—a newsletter of the IFLA Core Activity on Preservation and Conservation No. 59-60

Wang Z, Li X (2014) World cultural heritage site—the Eastern Qing Tombs cultural landscape—the Highest Expression of Chinese Fengshui theory in heritage and landscape as human values, ICOMOS, Florence. pp 60–67

Watson B (1993) Records of the Grand Historian: Qin Dynasty (trans: Qian S). Columbia University Press, New York

Wei H (2017) The evolution of auspicious beasts in the burial decoration of the Six-Dynasties Period and the formation of the "Jin System". Chin Archaeol 17:187–192

Wei C, Weixing Z (2015) Emperor Qin Shihuang's Mausoleum site in Xi'an. Chin Archaeol 15(1)

Wheatley P (1971) The pivot of the four quarters: a preliminary enquiry into the origins and character of the Ancient Chinese City. Edinburgh University Press, Edinburgh

Wheatley P (1975) The ancient Chinese city as a cosmological symbol. Ekistics 39(232):147–158

Whiteman S (2013) Kangxi's Auspicious Empire: Rhetorics of geographic integration in the Early Qing. In: Du Y, Jeffrey K-M (eds) Chinese history in geographical perspective. Lexington Books, Lanham

Wilkinson E (2000) Chinese history: a manual. Harvard University Press, Cambridge

Witten J, Hastie D, Tibshirani R (2013) An introduction to statistical learning with applications in R. Springer, New York

Wu H (1995) Monumentality in Early Chinese Art and Architecture. Stanford University Press, Stanford

Wu H (2009) Rethinking East Asian Tombs: a methodological proposal. In: Studies in the history of art, vol 74, Symposium papers: dialogues in art history, from Mesopotamian to Modern: Readings for a New Century, pp 138–165 National Gallery of Art, London

Wu H (2010) The art of the yellow springs: understanding Chinese Tombs Honolulu. University of Hawai'i Press

Wu J, Chen M, Liu C (2009) Astronomical function and date of the Taosi observatory. Sci China Ser G-Phys Mech Astron 52:151–158

Wuzhan Y (2011) Textual research on the sealing clays unearthed from the Yangling Mausoleum of the Western Han. Wenwu 04:1

Wuzhan Y (2017) Discussion on the function of Western-Han Mausoleum Towns. Wenwu 03

Xiuzhen L, Bevan A, Martinón-Torres M, Xia Y, Zhao K (2016) Marking practices and the making of the Qin Terracotta Army. J Anthropol Archaeol 42:169–183

Xu W (2015) The Excavation of the Terracotta Army Pit No. 1 of Emperor Qin Shihuang Mausoleum in 2009–2011. Wenwu 9:4–38

Xu Y (2019) Grids of Chinese ancient cities. Spatial planning tools for achieving social aims. Altralinea ed, Milano

Yi D, Yu L, Hong Y (1994) Chinese Traditional Feng Shui Theory and Building Site Selection. Science and Technology Press, Hebei

Yoon H (2008) Culture of Feng Shui in Korea: an exploration of East Asian Geomancy. Lexington Books, London

Yoshitaka H (2017) Burial mounds, volcanoes, and the sun. Quat Res (Daiyonki-Kenkyu) 56(3):97–110

Zhang J (2004) A Translation of the Ancient Chinese 'The Book of Burial (Zang Shu)' by Guo Pu (276–324). Edwin Mellen Press, New York

Zhang W (2016) Ritual and Order: Research on the Mausoleum of the First Qin Emperor. Science Publishing, Beijing

Zhewen L (1993) China's Imperial Tombs and Mausoleums. Foreign Languages Press, Beijing

Zhongshu Z (1992) Round Sky and Square Earth (Tian Yuan Di Fang): Ancient Chinese geographical thought and its influence. GeoJournal, vol 26, no 2. History of Geographical Thought

Zhou H (2009) Zhaoling: The Mausoleum of Emperor Tang Taizong. Victor H. Mair. Editor. Sino-Platonic Papers 187:1–380

Zhu J (2003) Chinese spatial strategies: imperial Beijing, 1420–1911 Routledge, London

Zubrow EB (1994) Knowledge representation and archaeology: a cognitive example using GIS. In: Renfrew C, Zubrow EBW (eds) The Ancient Mind. Elements of Cognitive Archaeology. Cambridge University Press, Cambridge

Printed in the United States
by Baker & Taylor Publisher Services